SPACE
RADIOBIOLOGY

Synergies between Astroparticle and Medical Physics

SPACE RADIOBIOLOGY

Synergies between Astroparticle and Medical Physics

Alessandro Bartoloni

Istituto Nazionale di Fisica Nucleare, Italy

Lidia Strigari

IRCCS Azienda Ospedaliero-Universitaria di Bologna, Italy

World Scientific

NEW JERSEY · LONDON · SINGAPORE · BEIJING · SHANGHAI · HONG KONG · TAIPEI · CHENNAI · TOKYO

Published by

World Scientific Publishing Europe Ltd.

57 Shelton Street, Covent Garden, London WC2H 9HE

Head office: 5 Toh Tuck Link, Singapore 596224

USA office: 27 Warren Street, Suite 401-402, Hackensack, NJ 07601

Library of Congress Cataloging-in-Publication Data

Names: Bartoloni, Alessandro author | Strigari, Lidia author

Title: Space radiobiology : synergies between astroparticle and medical physics /
   Alessandro Bartoloni, Istituto Nazionale di Fisica Nucleare, Italy;
   Lidia Strigari, IRCCS Azienda Ospedaliero-Universitaria di Bologna, Italy.

Other titles: Synergies between astroparticle and medical physics

Description: New Jersey : World Scientific, [2026] | Includes bibliographical references and index. |
   Contents: Space Radiation and Human Space Exploration -- Measuring Space Radiation --
   Monitoring Space Radiation -- Health Risks of Radiation -- Radiation Risk Assessment for
   human in space missions -- Ionizing Radiation Medical Applications --
   Unlocking Interdisciplinary Research : Synergies in Radiobiology.

Identifiers: LCCN 2025020926 | ISBN 9781800617674 hardcover |
   ISBN 9781800617681 ebook | ISBN 9781800617698 ebook others

Subjects: LCSH: Space radiobiology | Medical physics | Particles (Nuclear physics)

Classification: LCC QH328 .B37 2026

LC record available at https://lccn.loc.gov/2025020926

British Library Cataloguing-in-Publication Data

A catalogue record for this book is available from the British Library.

Cover image: The logo on the astronaut's arms was created with the assistance of AI image generation (OpenAI's DALL·E), blending elements of abstract symbolism with the RGB colour palette—red, green, and blue—to represent the unity of diverse nations and disciplines. The intertwining forms evoke orbital paths and shared human heritage, reflecting the hope for a future of peaceful co-operation and exploration beyond Earth.

For any available supplementary material, please visit
https://www.worldscientific.com/worldscibooks/10.1142/Q0521#t=suppl

Desk Editors: Murali Appadurai/Shi Ying Koe

Typeset by Stallion Press
Email: enquiries@stallionpress.com

To my dearest daughters, Marina and Milena,
You are my greatest joy and my constant inspiration.
With all my love,
Dad
— Alessandro Bartoloni

To my remarkable co-author,
Thank you for opening the door to this extraordinary world of
knowledge and inviting me on this incredible adventure.
Your passion, guidance, and unwavering belief in our
shared journey have been my greatest inspirations. This book
is a tribute to our collaboration, curiosity, and the paths
we dared to explore together.
— Lidia Strigari

About the Authors

Alessandro Bartoloni is a scientist at the Istituto Nazionale di Fisica Nucleare (INFN), Italy, and user associate at CERN. For 30 years, his research has focused on interdisciplinary fields, spanning supercomputing for breakthrough applications in physics, the development of innovative detectors for high-energy physics, and cosmic ray measurements in space. He has collaborated on international projects that have led to significant scientific discoveries, including the 2012 identification of the Higgs boson. Since 2001, he has been part of the Alpha Magnetic Spectrometer (AMS) collaboration. In 2017, as the leader of the AMS group at Sapienza University of Rome's INFN, he promoted and established an interdisciplinary research team that integrated physicians and medical physicists. This collaboration explores the intersection of astroparticle physics and medical physics, with a specific focus on space radiobiology. His research plays a crucial role in developing strategies to mitigate radiation risks in space exploration.

Lidia Strigari is the head of the Department of Medical Physics at the IRCCS Azienda Ospedaliero-Universitaria di Bologna, Italy. Her work focuses on medical physics in translational environments, particularly radiobiology, with applications in radiotherapy, theragnostics, and space exploration missions. She leads and collaborates on international projects, developing innovative strategies for radiation protection and personalized medicine. Additionally, she has contributed to numerous international committees and working groups, supported by her publication track record. Her interdisciplinary research enhances our understanding of radiation exposure, bridging fundamental physics with clinical applications to improve safety measures in both space exploration and medical treatments on Earth.

Acknowledgment by A. Bartoloni

I express my deepest gratitude to the AMS Collaboration I joined in 2001. Without it, this book would not have been possible.

I want to thank the INFN APE and CERN CMS Collaboration for shaping me as a scientist and the INFN Roma Sapienza division and astroparticle scientific community for their unwavering support in my research endeavors.

Thank you to all the friends and colleagues I have met over the past six years for your insightful discussions. In particular, I am deeply grateful to Marco Bochicchio, Bruno Borgia, and Gianluca Cavoto for inspiring the early steps and to Guenter Reitz and Livio Narici for inviting me to participate in the international space radiation community.

Thank you all for your contributions and encouragement on this journey.

Acknowledgment by L. Strigari

I would like to express my deepest gratitude to my beloved family for their unwavering support and encouragement throughout every challenge and milestone of my journey. Your love has been the foundation upon which I have built both my personal and professional life.

I am also sincerely grateful to my colleagues, whose professionalism and innovative spirit have continually inspired me. Your dedication has greatly contributed to my growth and development in the field of medical physics.

Contents

Acronyms

ACC	Anticoincidence Counter
ACE	Advanced Composition Explorer
ACR	Anomalous Cosmic Rays
ADC	Analog-to-Digital Converter
AGN	Active Galactic Nucleus
AI	Artificial Intelligence
ALTEA	Alpha, Lithium, Tritium Experiment Apparatus
AMS	Alpha Magnetic Spectrometer
AMS-100	Alpha Magnetic Spectrometer-100
AR	Active Region
ARS	Acute Radiation Syndrome
ART	Adaptive Radiotherapy
ASC	Active Scintillation Counter
ASI	Agenzia Spaziale Italiana
ASIC	Application-Specific Integrated Circuit
BER	Base Excision Repair
BFO	Blood Forming Organ
BLEO	Beyond Low Earth Orbit
BON	Badhwar–O'Neill Model
CAD	Computer-Active Dosimeter *(onboard ARTEMIS I)*
CALET	Calorimetric Electron Telescope

CBCT	Cone Beam Computed Tomography
CGBM	CALET Gamma-Ray Burst Monitor *(onboard CALET)*
CHD	Charge Detector *(onboard CALET)*
CME	Coronal Mass Ejection
CNS	Central Nervous System
CP	Charged Particle
CPD	Charged Particle Detector *(onboard RAD)*
CR	Cosmic Ray
CRATER	Cosmic Ray Telescope for the Effects of Radiation *(onboard LRO)*
CRD	Cosmic Ray Detector
CREM	Cosmic Radiation Environmental Model
CRÈME	Cosmic Ray Effects on Micro-Electronics
CRIS	Cosmic-Ray Isotope Spectrometer *(onboard ACE)*
CRRES	Combined Release and Radiation Effects Satellite
CSM	Command Service Module *(on Apollo missions)*
CSP	Cell Survival Probability
CSS	Chinese Space Station
CT	Computed Tomography
CTV	Clinical Target Volume
CVD	Cardiovascular Disease
CZT	Cadmium Zinc Tellurate
DAMPE	Dark Matter Particle Explorer
DBT	Digital Breast Tomosynthesis
DIS	Direct-Ion Storage Technology
DMSA	Dimercaptosuccinic Acid
DNA	Deoxyribonucleic Acid
DSA	Digital Subtraction Angiography
EAD	ESA Active Dosimeter *(onboard ARTEMIS I)*
EBIS	Electron Beam Ion Source *(in NSRL)*
ECAL	Electromagnetic Calorimeter
EHIS	Energetic Heavy Ion Sensor *(onboard SEISS)*

ELISA	Enzyme-Linked Immunosorbent Assay
EMIC	Electromagnetic Ion Cyclotron
EPAM	Electron, Proton, and Alpha-Particle Monitor *(onboard ACE)*
EPID	Electronic Portal Imaging Device
ESA	European Space Agency
FBP	Filtered Back Projection
FDG	Fluorodeoxyglucose
FET	Field-Effect Transistor
FISH	Fluorescence *In Situ* Hybridization
FLT	Fluorothymidine
FND	Fast Neutron Detector *(onboard ISS-RAD)*
GCR	Galactic Cosmic Ray
GOES	Geostationary Operational Environmental Satellite
GTV	Gross Tumor Volume
HALO	Habitation and Logistics Outpost *(ARTEMIS Moon Gateway)*
HDPE	High-Density Polyethylene
HDR	High-Dose-Rate
HED	High-Energy Density Region
HEO	Highly Elliptical Orbit
HEP	High Energy Physics
HERA	Hybrid Electronic Radiation Assessor *(onboard ARTEMIS I)*
HERD	High Energy Cosmic-Radiation Detection
HF	Hazard Function
HIMAC	Heavy Ion Medical Accelerator
HMPAO	Hexamethylpropyleneamine Oxime
HN	Heavy Nuclei
HR	Homologous Recombination
HRS	Hyper Radiosensitivity
IAF	International Astronautical Federation

ICRP	International Commission on Radiological Protection
IMF	Interplanetary Magnetic Field
IMRT	Intensity Modulated Radiation Therapy
INFN	Istituto Nazionale di Fisica Nucleare
IORT	Intra-Operative Radiation Therapy
IR	Ionizing Radiation
IRR	Increased Radio Resistance
ISECG	International Space Exploration Coordination Group
ISM	Interstellar Medium
ISS	International Space Station
ISS-CREAM	Cosmic Ray Energetics and Mass for the International Space Station
IVAB	Inner Van Allen Belt
JWST	James Webb Space Telescope
LAFOV	Large Area Field of View
LDR	Low-Dose-Rate
LEO	Low Earth Orbit
LET	Linear Energy Transfer
LIDAL	Light Ions Detector for ALTEA
LINAC	Linear Accelerator
LIS	Local Interstellar Spectrum
LM	Lunar Module *(on Apollo missions)*
LND	Lunar Lander Neutron and Dosimetry
LQ	Linear Quadratic
LQC	Linear Quadratic Cubic
LRO	Lunar Reconnaissance Orbiter
MAA	Macroaggregated Albumins
MAG	Magnetometer *(on ACE)*
MARE	Matroshka AstroRad Radiation Experiment *(onboard ARTEMIS I)*
MARIE	Mars Radiation Environment Experiment
MAVEN	Mars Atmosphere and Volatile Evolution

MC	Monte Carlo
MDC	Mission Data Controller *(onboard CALET)*
MDP	Methylene Diphosphonate
MDR	Maximum Detectable Rigidity
MEO	Medium Earth Orbit
MR	Magnetic Resonance
MRI	Magnetic Resonance Imaging
MRO	Mars Reconnaissance Orbiter
MVCT	Mega Voltage Computer Tomography
NASA	National Aeronautics and Space Administration
NCRP	National Council on Radiation Protection and Measurements
ND	Neutron Detector
NER	Nucleotide Excision Repair
NGS	Next-Generation Sequencing
NHEJ	Non-Homologous End Joining
NPDS	Nuclear Particle Detection System *(onboard Apollo mission)*
NSRL	National Space Radiation Laboratory
NTE	Non-Targeted Effects
OVAB	Outer Van Allen Belt
PAMELA	Payload for Antimatter Matter Exploration and Light-Nuclei Astrophysics
PBS	Pencil Beam Scanning
PCR	Primary Cosmic Rays
PDR	Pulsed-Dose-Rate
PET	Positron Emission Tomography
PPE	Personal Protective Equipment
PRD	Personal Radiation Dosimeter *(onboard the Apollo missions)*
PSD	Plastic Scintillator Detector
PSMA	Prostate-Specific Membrane Antigen

PTV	Planning Target Volume
QF	Quality Factor
RAD	Radiation Assessment Detector *(onboard the ISS)*
RAM	Radiation Area Monitor
RBE	Relative Biological Effectiveness
REID	Risk of Exposure-Induced Death
REM	Radiation Environment Monitor *(onboard the ISS)*
RICH	Ring Imaging Cherenkov
RMR	Repair Misrepair
RMS	Radiation Monitoring System *(onboard the ISS)*
ROS	Reactive Oxygen Species
RSM	Radiation Survey Meter *(onboard the Apollo missions)*
RT	Radiation Therapy
RTSW	Real-Time Solar Wind *(onboard ACE)*
SAA	South Atlantic Anomaly
SBRT	Stereotactic Body Radiation Therapy
SCD	Silicon Charge Detector *(onboard ISS-CRÈME)*
SCR	Secondary Cosmic Ray
SEISS	Space Environment *In Situ* Suite *(onboard GEOS satellites)*
SEP	Solar Energetic Particle
SF	Solar Flare
SHARP	Space-Weather HMI Activity Region Patches
SIB	Simultaneous Integrated Boost
SIS	Solar Isotope Spectrometer *(on ACE)*
SKY	Spectral Karyotyping
SLS	Space Launch System *(ARTEMIS program)*
SNR	Supernova Remnants
SOD	Superoxide Dismutase
SOFT	Scintillating Optical Fiber Trajectory *(onboard ACE)*
SPE	Solar Particle Event
SPECT	Single Photon Emission Computed Tomography

SR	Space Radiation
SRAG	Space Radiation Analysis Group
SSC	Space Station Computer *(onboard the ISS)*
SWICS	Solar Wind Ion Composition Spectrometer *(onboard ACE)*
SWIMS	Solar Wind Ion Mass Spectrometer *(onboard ACE)*
TEP	Tissue-Equivalent Plastic
TEPC	Tissue-Equivalent Proportional Counter
TGO	Trace Gas Orbiter
TLD	Thermoluminescent Dosimeter
TOF	Time of Flight
TP	Tumor Prevalence
TRD	Transition Radiation Detector
TST	Track Structure Theory
UHF	Ultrahigh Frequency
ULEIS	Ultralow Energy Isotope Spectrometer *(onboard ACE)*
ULF	Ultralow Frequency
UP	Uncharged Particles
VAB	Van Allen Belt
VMAT	Volumetric Modulated Arc Therapy

Chapter 1

Introduction

1.1 Introduction

Cosmic rays (CRs) approaching our planet interact with the Earth's magnetic field and atmosphere, which remove or deflect a significant number of CRs so that the average annual dose due to these radiations received by humans living on the Earth's surface is about 0.33 mSv, representing approximately 10% of the natural background radiation [1, 2]. The Earth's atmosphere acts like a metallic shield several meters thick against radiation. CR intensity at the surface varies with latitude, increasing both north and south of the equator because the Earth's magnetic field deflects high-velocity charged particles (CPs) crossing the magnetic force field. Moreover, CR intensity increases with altitude, as the atmosphere thins [3, 4].

The situation in space is completely different, where the CRs, if interacting with the human body, could release absorbed doses at potentially dangerous levels for human health. In this regard, space radiation is one of the main concerns for ensuring safe space exploration, as planned by all national space agencies in the coming years [5, 6]. In this context, all the different components of space radiation (SR) have been extensively studied and measured over the last decades by numerous astroparticle experiments operating in space, and the data obtained from these experiments can be used to improve radiation health risk assessment for humans on space missions. This could also improve our understanding of the effects of radiation quality, including the various quantitative and qualitative damages that each type of radiation can cause.

1.2 The Space Radiation Environment

The SR environment (Fig. 1.1) is a complex mixture of radiation species, both charged and uncharged, dominated by highly penetrating CPs from various sources. Space particles can be categorized based on their origin, starting from those that are farthest away, in the following way:

- *Galactic cosmic rays (GCRs)*, which originate outside the solar system. GCRs are accelerated by sources in our galaxy or produced as secondary particles through the interaction of primary accelerated particles with the interstellar medium. GCRs consist of CPs (such as protons and ions) and neutral particles. The neutral components of GCRs are typically composed of photons, neutrons, and neutrinos. Therefore, these neutral particles do not carry an electric charge and are unaffected by space electromagnetic fields. They can travel through space relatively unaffected by magnetic fields and penetrate matter more efficiently than CPs.
- *Sun radiation*, or *solar energetic particles (SEPs)*, which are accelerated by the Sun during explosive events such as solar flares (SFs) and coronal mass ejections (CMEs).
- *Trapped radiation* particles created within the Earth's magnetosphere are of secondary origin. When a primary particle from outside the magnetosphere interacts with the atoms of the Earth's mesosphere, secondary particles are created and trapped in the radiation belts.

The so-called solar modulation of CRs is also relevant in the heliosphere. It refers to the influence of the Sun's magnetic field and solar wind on the flux and energy spectrum of CRs in the solar system. The solar wind, a stream of CPs emitted by the Sun, carries the Sun's magnetic field throughout the solar system. CRs propagate through the heliosphere and interact with the solar wind and its magnetic field. This interaction leads to the modulation of CRs, affecting their flux and energy distribution. The modulation process is a result of several factors. First, the expanding solar wind and its embedded magnetic field act as a shield, reducing the flux of CRs within the heliosphere compared to their flux in interstellar space. This reduction in flux is more pronounced for low-energy CRs due to their greater susceptibility to the effects of solar wind. Second, the strength and

Galactic Cosmic Ray
- Origin: outside the solar system
- Composition:
  - Electrons: ~1%
  - Protons: ~88%
  - Helium: ~10%
  - Heavier Ions: ~1%

Sun Radiation
- Origin: The Sun produces radiation in the form of solar energetic particles (SEPs) and solar wind
- Composition:
  - Protons: ~95% of SEPs
  - Helium: ~4%
  - Heavier ions (e.g., oxygen, iron): ~1%

Van Allen Belts
- Origin: Trapped radiation in the magnetosphere mainly concentrated in two areas
  - **Inner belt:** ~0.2 to 8 km from the Earth's surface
  - **Outer belt:** ~15,000 to 65,000 km from the Earth's surface
- Composition:
  - Inner belt: Predominantly high-energy protons and electrons
  - Outer belt: Primarily energetic electrons

Fig. 1.1. The space radiation environment is a complex mixture of ionizing radiations that originate outside the solar system, are generated by the Sun, and are trapped in the magnetosphere.

Source: Generated with licensed MS Copilot tool by the authors.

configuration of the Sun's magnetic field vary over time due to the solar activity cycle. The solar activity cycle, characterized by variations in the number of sunspots and the polarity of the Sun's magnetic field, has an approximately 11-year periodicity known as the solar cycle. During periods of high solar activity, such as solar maximum, the Sun exhibits more frequent explosive events, including SFs and CMEs. These events can generate enhanced SEPs with a temporary increase in the flux of energetic particles, including CRs. The occurrence of SEPs can further modulate the CR flux near the Earth and other planets. The solar modulation effect is essential for space missions involving human astronauts. The flux and energy distribution of CRs experienced by astronauts during their space journeys is influenced by solar activity. Higher solar activity, associated with increased solar particle events, can pose radiation risks to astronauts. Therefore, understanding and predicting solar modulation patterns are crucial for planning space missions and ensuring the safety of astronauts in space.

On space stations or spacecraft, SR environments are also influenced by shielding, which modifies the incident spectrum and related exposure due to the attenuation of the energy of incident particles and the production of new particles by interaction (spallation) with such structures. Such particles, often referred to as secondary, can penetrate several tens of centimeters of materials such as aluminum or tissue/water and can pose a greater health risk to astronauts than primary particles.

Similarly, the Earth's magnetic field acts as a barrier for incoming particles, with permitted and forbidden trajectories depending on the particles' energy and incoming direction, i.e., the angle between the particles' trajectory and the magnetic field (the pitch angle). So, the particle spectrum measured inside the Earth's magnetosphere differs from that of the outside.

Radioprotection science is the study of technical solutions applied to reduce exposure for both the general public and field workers, with criteria generally based on three principles [7]:

- increasing the distance from the radiation source,
- reducing exposure time,
- implementing *ad hoc* shielding.

Distance is not helpful in space, as GCRs are substantially isotopically distributed. Time in space is carefully managed to minimize duration according to exploration plans or by decreasing flight time. Therefore, shielding materials cannot fully absorb all SRs due to the very high-energy component of the GCR spectrum. In addition, shielding needs to be optimized, considering its efficacy and cost, to reduce unavoidable exposures to the minimum acceptable level. More in detail, passive or active shielding may significantly reduce radiation exposure, considering the time-varying contribution of GCRs. Still, passive shielding is known, and has been known for nearly five decades, to be largely ineffective at significantly reducing exposure from GCRs. Active shielding technologies have been proposed multiple times over the past few decades, but not a single approach has proved capable of markedly reducing exposure from energetic ions. The severe practical engineering constraints that hinder active shielding tech-nologies have also not been overcome. The expected exposure in beyond low Earth orbit (BLEO) exploration is significantly higher, even hundreds of times, than on the Earth's surface. This could imply that astronauts may exceed the recommended lifetime radiation exposure limit [8].

Table 1.1 presents estimated exposure doses for astronauts in various space mission scenarios, along with their corresponding ratios relative to Earth's surface exposure [9]. The absorbed dose quantifies the energy deposited by radiation per unit mass of tissue, while the equivalent dose adjusts for the type of radiation and its biological effects. The effective dose further accounts for the varying radiosensitivity of different organs. For a more detailed discussion, refer to Section 6.3.

In this context, the developments and continuous improvements of risk models allow for accurate estimation of the IR effects on the health of space travelers. The most recent approaches are based on different quantities and concepts and can be summarized as follows [10]:

- *Particle quality factor (QF) and radiation weighting factor (WR)*: The QF and WR values are used to quantify the relative biological effectiveness (RBE) of different types of radiation. These factors have undergone refinements based on new experimental data and epide-miological studies, leading to more accurate assessments of the risks associated with different radiation types.
- *RBE modeling*: RBE models aim to improve our understanding of the biological effects of different types of radiation in space. They con-sider radiation types, energy, dose, dose rate, and biological endpoints to estimate the relative risks of various radiation exposures.
- *Linear energy transfer (LET)*: Updated risk models also account for the varying levels of LET associated with different types of radiation. LET plays a crucial role in assessing the biological effects of radiation exposure, and recent models consider the LET distribution and its impact on cellular and tissue damage.
- *Integrated risk models*: Contemporary risk models aim to integrate various factors, including dose, radiation quality, and individual

Table 1.1. Daily absorbed, equivalent, and effective ionizing radiation doses in different exposure scenarios. Also, some ionizing radiation doses typical of clinical use are reported.

Exposure Scenario	Ratio	Daily Absorbed Dose (mGy)	Daily Equivalent Dose (mSv)	Daily Effective Dose (mSv)
Earth's Surface (Background Radiation)	1.00		0.008	
Earth's Surface (Cosmic Radiation Only)	0.13		0.001	
LEO-ISS (Average)	51.37	0.41	0.41	0.41
Mars's Surface (Solar Maximum)	37.50	0.12	0.30	0.28
Mars's Surface (Solar Minimum)	76.25	0.25	0.61	0.54
Lunar Surface (Solar Maximum)	72.50	0.10	0.58	0.38
Lunar Surface (Solar Minimum)	195.00	0.27	1.56	0.84
Typical Radiation Workers' Exposure	**Ratio**	**Absorbed Dose (mGy)**	**Equivalent Dose (mSv)**	**Effective Dose (mSv)**
Aircrew Flying Polar Routes (Per Year)	2.05		6.00	
Diagnostic Radiologist (Per Year)				0.9 [6]
Clinical Examples (Diagnosis)	**Ratio**	**Absorbed Dose (mGy)**	**Equivalent Dose (mSv)**	**Effective Dose (mSv)**
Whole-Body CT (Per Exam)	482.00			4.82
PET/CT (Per Exam)	1840.00			18.4
Cardiac SPECT/CT (Per Exam)	1270.00			12.7
Clinical Examples (Treatment)	**Ratio**	**Absorbed Dose (mGy)**	**Equivalent Dose (mSv)**	**Effective Dose (mSv)**
Proton Therapy Dose/Fraction (Typical)	220000.00		2200	

Note: The "Ratio" column is calculated with respect to Earth's surface exposure, using the equivalent dose value where specified and the equivalent dose estimated from other cases where not specified.

susceptibility, to provide a comprehensive assessment of radiation risks in space. These models consider the combined effects of different radiation types and exposure scenarios to estimate the overall risk to astronauts more effectively.

1.3 Space Radiation Interaction with Matter

SR includes both CPs and uncharged particles (UPs). Secondary particles produced during the collision of space particles with matter are direct consequences of all the different interaction types occurring, as summarized in Table 1.2 [11–14]. The table distinguishes between CPs, such as electrons, protons, and heavy ions, and UPs, such as neutrons and photons. Each interaction type (e.g., Coulomb interaction, Bremsstrahlung, and spallation) is listed along with its corresponding primary particles.

Table 1.2 demonstrates the complexity of the SR environment.

Table 1.2. Overview of the interactions of space radiation with matter and the resulting secondary particles.

Interaction Type	Primary Particle	Secondary Particles
Coulomb Interactions	Charged particles (CPs)	Electrons, ions, and photons (via ionization and excitation)
Bremsstrahlung Emission	CPs	Photons
Nuclear Elastic and Inelastic Collisions	CPs and Uncharged particles (UPs)	Secondary neutrons, protons, heavy ions, and gamma rays
Pair Production	High-energy photons	Electron–positron pairs
Photoelectric Absorption	Photons	Electrons
Compton Scattering	Photons	Electrons and lower-energy photons
Neutron Interactions	Neutrons	Gamma rays, protons, alpha particles, deuterons, and additional neutrons
Spallation	High-energy CPs or Neutrons	Protons, neutrons, heavy ions, and fragments of nuclei
Cherenkov Radiation	High-speed CPs	Photons (optical and ultraviolet light)

1.4 Particle Interaction with Cells and Tissues

CPs and UPs can cause damage to human organs and tissues through their interactions with biological materials. The specific damage to organs and tissues depends on factors such as the energy, dose, duration of exposure, and the sensitivity of different tissues to radiation. Shielding and proper radiation protection measures are essential to minimize exposure to uncharged particles and reduce the potential damage to human organs and tissues in radiation environments. Biophysical models are crucial in understanding the interaction mechanism between radiation and organisms. Many radiobiological models have been proposed to explain the survival fraction and estimate the RBE in clinical applications (mainly radiotherapy, where irradiated tissues are subjected to high doses per fraction). In contrast, SR radiobiological models focus on low-dose effects, i.e., they investigate the RBE associated with tumorigenesis or related surrogate endpoints (such as chromosome aberrations or mutations). Nevertheless, an accurate estimation of the radiation QF might improve the models proposed for SR risk assessment in interplanetary missions [15].

Regarding UPs, neutrons are particularly concerning because they have no electric charge and can penetrate deep into tissues, leading to ionizations and nuclear interactions. Neutrons have a higher relative RBE than photons, meaning that they are more effective at causing biological damage for the same absorbed dose. Their interactions with atoms in the body can produce secondary charged particles, further increasing the potential for biological damage.

GRs are high-energy photons and can also penetrate tissues, causing ionization along their path. They are highly penetrative and can interact with many atoms in the body, producing ionized atoms and free radicals. This can result in DNA damage, cell death, and disruption of normal cellular functions. The severity of the damage depends on the energy and dose of the gamma rays. Both neutrons and gamma rays can cause a range of acute and long-term effects on human health, including radiation sickness, cancer, and genetic mutations.

CPs penetrating living matter possess physical and radiobiological properties, including a sharp ionization maximum near the end of their penetration range (i.e., the Bragg peak). This implies that the effects indicated by the RBE quantities are enhanced compared to those of reference radiation (usually X-rays or GRs). In other words, absorbed doses lower than those released by the reference radiation will produce an equivalent biological effect for a given experimental observation. Indeed, the complex RBE depends not only on dose levels but also on other factors (such as the dose rate and the LET) and biological characteristics (such as species, tissues, and the endpoint under consideration). RBE measures the radiation quality used to induce a given endpoint. In addition, the extent of damage to human organs and tissues depends on the radiosensitivity of cells and the functional architecture of organs and tissues.

1.5 Health Hazards of Space Radiation to Astronauts

SR presents significant health risks to astronauts, leading to acute and long-term effects. Acute radiation syndrome (ARS) arises from brief, intense exposure to SEPs. In contrast, prolonged exposure to GCRs and SFs poses severe long-term dangers. These radiation-induced health hazards can result in life-threatening conditions, including cancer, neurological disorders, and cardiovascular damage. Understanding and mitigating these effects is crucial for ensuring astronaut safety during extended space missions. The following points outline key challenges and research directions in this field.

- *Acute radiation syndrome*: ARS occurs when astronauts experience intense, short-term exposure to SEPs without adequate shielding. The Sun releases high-energy particles, mainly protons and electrons, which can penetrate spacesuits and spacecraft shielding, leading to harmful radiation exposure.
  Symptoms of ARS include nausea, vomiting, diarrhea, fatigue, hair loss, and skin burns. Severe cases can cause damage to the bone marrow, gastrointestinal system, and central nervous system. The severity and onset of symptoms depend on radiation dose, exposure duration, and individual susceptibility. Prompt medical intervention and continuous monitoring are essential for managing ARS and minimizing its long-term effects.

- *Late effects of space radiation*: GCRs and SFs present distinct radiation risks due to their differing intensities and exposure patterns:
  - *GCRs*: These high-energy particles originate outside the solar system and maintain a relatively constant intensity. Although they are unlikely to cause ARS, prolonged exposure increases the risk of chronic conditions.
  - *SFs*: These sudden bursts of solar radiation can result in significantly higher short-term dose rates, potentially leading to ARS if an astronaut is exposed without adequate shielding.

  Long-term exposure to GCRs is linked to late radiation effects, including cancer, central nervous system damage, and cardiovascular diseases. The impact of GCRs varies based on factors such as solar activity, as discussed in Section 1.1. Additionally, ionizing radiation (IR) may contribute to secondary cancers and other health complications. Shielding and mitigation strategies are essential for reducing these risks in space missions.
- *Radiosensitivity and individual risk assessment*: Understanding cellular radiosensitivity is critical for identifying individual variations in radiation response [16]. Assessing astronauts' radiation resistance can help predict both early and late effects, allowing for personalized medical strategies. Mission crew selection should incorporate medical assessments of each member's radiosensitivity, complemented by onboard biodosimetry tools for precise diagnostics and in-flight care.
- *Dose–effect relationships and models*: Various dose–effect models have been developed to evaluate the biological impact of SR. These models predict clinical and subclinical effects on astronauts, drawing insights from various sources:
  - *In vitro studies*: examining molecular and cellular damage in human cells exposed to radiation.
  - *In vivo studies*: observing radiation effects on human subjects or animal models under controlled conditions.

  These studies contribute to our understanding of radiation-induced health risks during deep-space missions, enabling the development of better protection strategies and safety protocols.
- *Clinical diagnostic and radiotherapeutic simulations*: Ground-based clinical diagnostic and radiotherapeutic devices valuable tools to simulate SR exposure. They include technologies such as the following:
  - Linear accelerators (LINACs) generate high-energy particle beams (photons, electrons, and protons) to replicate SR conditions.

○ Computed tomography (CT) scanners use X-rays to analyze biological responses to radiation exposure.

These devices allow researchers to investigate acute, late, and long-term effects, thereby refining radiobiological models for SR risk assessment. Although they cannot fully replicate the complexity of GCR, they provide essential data for improving astronaut health protection strategies.

- *Non-targeted effects of space radiation*: The interaction of SPs with healthy human tissues can induce non-targeted effects [17], which must be thoroughly understood to assess potential long-term health risks.

- *Dose equivalents in space radiation exposure*: Accurate calculation of dose equivalents requires consideration of the QFs and RBE of high-LET particle distributions. These parameters should be measured in actual SR environments to refine risk assessments and protective strategies [18].

- *Innovative radioprotection approaches and strategies*: A novel approach to astronaut protection involves inducing a hibernation-like state to mitigate radiation damage. Hypothermia has demonstrated radioprotective effects and is currently being explored as a potential countermeasure for long-duration missions [19].

Effective radiation protection and mitigation strategies are essential for ensuring astronaut safety. These include the following:

○ Monitoring solar activity: real-time tracking and forecasting of SEP events to provide early warnings.

○ Advanced shielding materials: developing spacecraft designs that minimize radiation exposure.

○ Pre-flight training and awareness programs: educating astronauts on ARS symptoms and protective measures during SF events.

○ Improved radiation monitoring technologies: enhancing real-time detection and response capabilities during space missions.

- *Biodosimetry and biomarkers*: Developing dose–effect models is critical for implementing precise countermeasures, such as localized shielding and advanced dosimetry. Biodosimetry is an emerging field with significant promise for real-time radiation exposure assessment and personalized astronaut protection [20].

- *Synergy with astroparticle experiments*: Astroparticle experiments have collected a vast number of datasets that can advance SR research.

These data contribute to addressing key questions in radiation protection, including:

- dose–response relationships for cancer and non-cancer risks,
- potential radiation dose thresholds for different biological systems,
- more accurate definitions of radiation quality in triggering biological responses.

Leveraging astroparticle research results can significantly enhance our understanding of SR hazards and improve astronaut safety for future deep-space exploration missions.

All the points briefly described above will be discussed in greater depth in the book, paving the way for a new era of theoretical and experimental approaches to addressing key open questions in radiation protection, such as the dose–effect relationship for cancer and non-cancer risk, the possible existence of dose thresholds and endpoints for different biological systems, and more accurate definitions of the impact of radiation quality in triggering biological response, thereby shedding light on one of the dominant aspects of SR.

References

[1] Leroy, C., and Rancoita, P. G. (2016). *Principles of Radiation Interaction in Matter and Detection* (4th ed.). Singapore: World Scientific Publishing, Singapore. https://doi.org/10.1142/9167.

[2] National Council on Radiation Protection and Measurements (NCRP) Report. (2009). Ionizing Radiation Exposure of the Population of the United States, NCRP Report No. 160. https://ncrponline.org/publications/reports/ncrp-report-160-2/.

[3] Fujitaka, K. (2005). High-level doses brought by cosmic rays. *International Congress Series*, 1276. https://doi.org/10.1016/j.ics.2004.11.045.

[4] Cember, H., and Johnson, T. E. (2009). *Introduction to Health Physics* (4th ed.). New York, NY: McGraw-Hill.

[5] Walsh, L., Schneider, U., Fogtman, A., Kausch, C., McKenna-Lawlor, S., Narici, L., Ngo-Anh, J., Reitz, G., Sabatier, L., Santin, G., *et al.* (2019). Research plans in Europe for radiation health hazard assessment in exploratory space missions. *Life Sciences in Space Research*, 21, 73–82. https://doi.org/10.1016/j.lssr.2019.04.002.

[6] The International Space Exploration Coordination Group (ISECG). (2022). Global Exploration Roadmap Supplement – Lunar Surface Exploration Scenario Update 2022. https://www.globalspaceexploration.org/?p=1184.

[7] Durante, M. (2014). Space radiation protection: Destination mars. *Life Sciences in Space Research*, 1, 2–9. https://doi.org/10.1016/j.lssr.2014.01.002.

[8] Zeitlin, C., Hassler, D. M., Cucinotta, F. A., Ehresmann, B., Wimmer-Schweingruber, R. F., Brinza, D. E., Kang S., Weigle, G., Böttcher, S., Böhm, E., *et al.* (2013). Measurements of energetic particle radiation in transit to Mars on the Mars science laboratory. *Science*, 340, 1080–1084. https://doi.org/10.1126/science.1235989.

[9] Valinia, A., John, R. A., David, R. F., Joseph, I. M., Jonathan, A. P., and Alonso, H. V. (2022). Safe Human Expeditions Beyond Low Earth Orbit (LEO). NASA/TM-20220002905 NESC-RP-20-01589.

[10] Cucinotta, F. A., Kim, M. H. Y., Chappell, L. J., Huff, J. L., Howerton, J. P., and Plante, I. J. (2016). Space radiation cancer risk projections and uncertainties – 2016. NASA Technical Paper TP-2016-218759. Retrieved from https://ntrs.nasa.gov/citations/20160007094.

[11] Evans, R. D. (1955). *The Atomic Nucleus*. New York, NY: McGraw-Hill.

[12] Meyerhof, W. E. (1967). *Elements of Nuclear Physics*. New York, NY: McGraw-Hill.

[13] Tsoulfanidis, N. and Landsberger, S. (2015). *Measurement and Detection of Radiation*. Boca Raton, FL: CRC Press.

[14] Heilbronn, L. and Nakamura, T. (2005). *Handbook on Secondary Particle Production and Transport by High-energy Heavy Ions*. Singapore: World Scientific Pub Co Inc.

[15] Dietze, G., Bartlett, D. T., Cool, D. A., Cucinotta, F. A., Jia, X., McAulay, I. R., Pelliccioni, M., Petrov, V., Reitz, G., and Sato, T. (2013). ICRP Publication 123: Assessment of radiation exposure of astronauts in space. *Annals of the ICRP*, 42, 1–339. https://doi.org/10.1016/j.icrp.2013.05.004.

[16] Quintens, R., Baatout, S., and Moreels M. (2019). Assessment of radiosensitivity and biomonitoring of exposure to space radiation. *Stress Challenges, and Immunity in Space*, pp. 519–533. Cham, Switzerland: Springer.

[17] Nelson, G. A. (2016). Space radiation and human exposures, a primer. *Radiation Research*, 185, 349–358. https://doi.org/10.1667/RR14311.1.

[18] Furukawa, S., Nagamatsu, A., Nenoi, M., Fujimori, A., Kakinuma, S., Katsube, T., Wang, B., Tsuruoka, C., Shirai, T., Nakamura, A. J., *et al.* (2020). Space radiation biology for "living in space." *BioMed Research International*, 2020, 25 p. https://doi.org/10.1155/2020/4703286.

[19] Cerri, M., Tinganelli, W., Negrini, M., Helm, A., Scifoni, E., Tommasino, F., Sioli, M., Zoccoli, A., and Durante, M. (2016). Hibernation for space travel: Impact on radioprotection. *Life Sciences in Space Research*, 11, 1–9. https://doi.org/10.1016/j.lssr.2016.09.001.

[20] Ainsbury, E. A., Moquet, J., Sun, M., Barnard, S., Ellender, M., and Lloyd, D. (2022). The future of biological dosimetry in mass casualty radiation emergency response, personalized radiation risk estimation, and space radiation protection. *International Journal of Radiation Biology*, 98, 421–427. https://doi.org/10.1080/09553002.2021.1980629.

Chapter 2

Space Radiation and Human Space Exploration

2.1 Introduction

Space radiation (SR) significantly differs from the types of radiation experienced on Earth's surface. In space, astronauts are exposed to highly hazardous charged particles (CPs) and stripped atomic nuclei, from which electrons have been removed, leaving only the atomic core. SR consists of three primary sources: particles trapped within Earth's magnetic field, particles emitted by the Sun during solar flares (SFs) and other solar events, and galactic cosmic rays (GCRs) originating from powerful astrophysical sources outside our solar system. The radiation environment in space includes a diverse range of particles, from light elementary particles such as electrons and protons to heavier nuclei, classified into particles with low ($Z < 8$), medium ($8 < Z < 14$), or high ($Z > 14$) nuclear charge. These unique radiation fields present conditions that differ starkly from any radiation exposure scenario encountered on Earth. Tables 2.1(a) and 2.1(b) report the estimated abundances of CPs in the near-Earth space region, where protons and helium nuclei constitute the majority of particles present. The abundances are given in ranges because cosmic-ray (CR) measurements are influenced by solar activity, which can temporarily alter the flux of different particle types reaching Earth. Additionally, the energy spectrum of CRs affects detection sensitivity for certain particles, leading to slight variations in measured abundances over time and across different observational instruments.

Table 2.1.   (a) Composition of cosmic rays near Earth by particle type. (b) A renormalized abundance of heavier nuclei ($Z > 2$) in cosmic rays near Earth.

(a)		
Particle Type	**Charge**	**Estimated Abundance (%)**
Electrons	−1	~1–2
Protons	1	~85–90
Helium Nuclei	2	~9–12
Heavier Nuclei ($Z > 2$)	>2	~1–3
(b)		
Nuclei Type	**Charge**	**Renormalized Abundance (%)**
Carbon	6	~20
Nitrogen	7	~3
Oxygen	8	~30
Neon	10	~6
Magnesium	12	~5
Aluminum	13	~2
Silicon	14	~5
Iron	26	~12
Others		~17

Note: The table shows the estimated abundance of each particle type as a percentage of the total number of CR particles detected near Earth. CRs predominantly consist of protons (hydrogen nuclei) and helium nuclei, with smaller fractions of electrons and heavier nuclei. The "Heavier nuclei" category ($Z > 2$) includes a variety of elements (e.g., carbon, oxygen, neon, silicon, and iron) that are relatively rare compared to lighter particles. Particle Data Group (PDG) Review of Cosmic Rays.

The table lists the most abundant nuclei types detected, along with their nuclear charge (Z) and estimated percentage abundances, normalized to sum to 100% within the heavier nuclei group. Oxygen and carbon are the most common heavier nuclei, followed by neon, magnesium, and iron. The "Others" category includes trace elements (e.g., lithium, beryllium, boron, fluorine, and sodium), with individual abundances below 1%. These values reflect typical CR compositions observed near Earth, where heavier nuclei comprise only a small fraction of the total particle population, which is dominated primarily by protons and helium nuclei. For astronauts venturing beyond Earth's protective atmosphere, one of the most significant challenges is managing exposure to high-energy radiation from the solar wind, solar storms, and GCRs. Beyond low Earth orbit (BLEO), SR poses substantial health risks, including acute radiation sickness, an elevated lifetime risk of cancer, impacts on the central nervous system, and the potential for degenerative diseases. Extensive research on radiation exposure has provided strong evidence that GCRs and solar particle events (SPEs) are associated with long-term health risks, including cancer and degenerative diseases. This chapter provides a comprehensive overview of the various components defining radiation fields and environments pertinent to upcoming space exploration missions within our solar system. It addresses the critical radiation protection challenges and strategies for diverse space exploration scenarios. These range from designing personal protective equipment (PPE) for individual astronauts and space workers to developing shielding solutions for different types of spacecraft. Additionally, the chapter examines protective strategies for habitats on the Moon, Mars, and in cis-lunar and deep-space environments. The objective is to highlight the key challenges and solutions necessary for protecting human health and ensuring mission success as we venture further into the demanding environments of deep-space exploration.

2.2 Sources of Space Radiation

Even though the primary aim of this chapter is to introduce the composition of SR and systematically highlight its potential risks, primarily defined by the charge, atomic number, and energy associated with CR particles, we also categorize SR by its sources. To illustrate, the impact of a proton with an energy of 100 MeV poses the same potential radiobiological damage, regardless of the astrophysical process that generated it. However, identifying the sources of SR is essential to understanding when and where it may pose the most significant risk. Accordingly, the following paragraph introduces the different components of SR within the heliosphere, classified by origin: GCRs, solar-generated particles, and those trapped within Earth's magnetosphere.

2.2.1 *Galactic cosmic rays*

GCRs represent the portion of CRs found in the solar system that originate from sources outside it and often from within the galaxy itself [1]. Various

Table 2.2.   Classification of critical cosmic ray isotopes as primary or secondary particles and their primary astrophysical sources and production mechanisms.

Nuclei Type	Astrophysical Source	Production Mechanism
Primary GCRs		
Hydrogen, Helium	SNRs, AGNs	Direct acceleration by shock waves
Carbon	SNRs, AGNs	Stellar nucleosynthesis, shock wave acceleration
Oxygen, Neon, Magnesium, Iron	SNRs	Stellar nucleosynthesis, shock wave acceleration
Secondary GCRs		
Lithium (^{6}Li, ^{7}Li), Beryllium (^{9}Be, ^{10}Be), Boron (^{10}B, ^{11}B)	ISM	Spallation of C, N, and O nuclei in ISM

astrophysical sources, such as active galactic nuclei (AGNs), supernova remnants (SNRs), and other high-energy phenomena, accelerate particles to form primary cosmic rays (PCRs). These are distinguished from secondary cosmic rays (SCRs), which are generated through interactions of PCRs with the interstellar medium (ISM) or other environments they traverse, such as planetary atmosoheres (see Table 2.2 for the classification of critical cosmic ray isotopes and their respective sources and production mechanisms). The ISM is a complex environment consisting of diffuse gases (mainly hydrogen and helium), dust, magnetic fields, and CRs that fill the space between stars. This medium is organized into phases based on temperature and density, ranging from cold, dense molecular clouds to hot, ionized gas. The ISM acts as both a target and a medium for CRs, where interactions between PCRs and ISM particles produce SCRs and new, often rare, isotopes. These interactions primarily occur through spallation, in which high-energy primary particles collide with nuclei in the ISM, breaking them into lighter particles and producing new elements and isotopes.

PCRs originate from energetic astrophysical sources and are accelerated to high energies by shock waves and magnetic fields associated with these events. These CRs consist of protons, helium, alpha particles, carbon, oxygen, neon, magnesium, silicon, and iron. These PCRs are accelerated by sources such as the following:

- *SNRs*: Shock waves from supernova explosions accelerate particles to high energies, making SNRs one of the most relevant sources of GCRs.

- *AGNs*: Supermassive black holes at the centers of galaxies produce jets and outflows, which can accelerate particles to relativistic speeds.
- *Pulsars and magnetars*: These neutron stars' intense magnetic fields and rapid rotations can accelerate particles and contribute to CR populations.

On the other hand, SCRs result from the interactions of PCRs with particles in the ISM. These interactions produce new isotopes and particle types, primarily lithium, beryllium, and boron. While these secondary elements are rare in PCRs, they are generated through a process known as spallation. Spallation occurs when high-energy PCRs, such as protons and heavy nuclei (e.g., carbon, nitrogen, and oxygen), collide with the nuclei of lighter elements, predominantly interstellar hydrogen and helium. During these collisions, the kinetic energy of the incoming CRs may be sufficient to overcome the binding energy of the target nucleus, leading to a nuclear reaction.

In a spallation reaction, the PCR imparts enough energy to eject one or more nucleons (protons or neutrons) from the target nucleus. The exact outcome of this process depends on several factors, including the energy of the incoming particle, the type of target nucleus, and the collision angle. The energy thresholds for spallation vary, but they are typically in the tens of MeV to hundreds of MeV.

When a CR nucleus collides with a target nucleus, several physical phenomena can occur:

- *Elastic scattering*: In some cases, the collision results in elastic scattering, where the two nuclei bounce off each other without any nucleons being ejected. This process does not contribute to spallation, but it can alter the trajectory of the CR.
- *Inelastic scattering*: More commonly, inelastic scattering occurs, leading to the disintegration of the target nucleus. This is the primary mechanism of spallation, where the incoming CRs transfer energy to the target nuclei, causing them to emit nucleons or lighter nuclei, such as protons or alpha particles.
- *Fragmentation*: The products of the spallation process can include lighter isotopes, such as lithium-6, lithium-7, beryllium-9, and boron-10. The emitted nuclei can also lead to further reactions with other nearby nuclei in the ISM, creating a chain of interactions that enhances the production of these secondary elements.

- *Energy loss and secondary production*: As PCRs travel through the ISM, they continuously lose energy through spallation and other interactions, producing a spectrum of secondary particles. This energy loss and the spallation process contribute significantly to the observed abundances of lithium, beryllium, and boron, which are not produced in significant quantities during stellar nucleosynthesis.

Spallation thus contributes significantly to the observed abundances of lithium, beryllium, and boron, which are otherwise not produced in large quantities by stellar processes.

PCRs originate directly from astrophysical events, such as SNRs and AGNs, where shock waves and magnetic fields accelerate particles to high energies.

Numerous astroparticle experiments on Earth and in space aim to measure CRs directly. These measurements can address unsolved fundamental physics questions, including the composition of dark matter and the possible existence of primordial antimatter.

CRs span a vast range of energies, from a few keV to well beyond TeV and even into the PeV range. The energy spectrum of CRs is typically represented as the flux (number of particles per unit area, time, and energy) as a function of energy, and it exhibits a strikingly smooth but steeply decreasing power-law distribution. This distribution means lower-energy CRs are far more abundant than high-energy ones, with the flux dropping sharply as energy increases (see Table 2.3 for a summary of cosmic ray flux characteristics across different energy ranges, including key spectral features and dominant sources).

The spectrum is a composite of all particles and varies by particle type. Each type of particle has its energy spectrum; however, generally, they follow a similar power-law shape with some notable features at certain energy thresholds. The proton and light nucleus spectra exhibit a steep decline with energy, but they have higher fluxes than those of heavier nuclei at a given energy. Heavy nuclei particles are less abundant overall, but they can reach incredibly high energies and contribute uniquely to the high-energy end of the spectrum. Electrons and positrons comprise a smaller portion of CRs but show distinctive features, especially at lower energies, influenced by both galactic sources and interactions with the ISM.

Describing the CR spectra, the concepts of "knees" and "ankles" are relevant; as energy increases, the spectrum of CRs displays, in fact, characteristic knee and ankle structures, transitions that appear as bends or

Table 2.3. This table summarizes the characteristics of cosmic ray flux across distinct energy ranges, including key spectral features, dominant sources, and approximate all-particle flux values. Approximate flux values are given as orders of magnitude.

CRs Energy Range	Typical Energy	Features	Source/Process Characteristics	Flux Particles/m²/sr/s/GeV
Low	~keV–GeV	Steep power-law decline	Galactic sources, such as solar flares and supernova remnants, with solar modulation affecting the flux	$\sim 10^4$–10^3
Intermediate	~GeV–1 PeV	Smooth spectrum, no major breaks	Primarily galactic sources with a steady power-law decay include protons, helium, and light nuclei	$\sim 10^3$–10^{-1}
First Knee (Galactic Transition)	~1–3 PeV	Spectrum steepens	Galactic sources (e.g., supernova remnants) reach maximum confinement energies, indicating acceleration limits in the galaxy	$\sim 10^{-1}$–10^{-4}
Second Knee (Heavy Nuclei)	~100 PeV	Further steepening of the spectrum	Likely associated with the maximum energy of heavy nuclei, such as iron, marking a limit for galactic sources	$\sim 10^{-4}$–10^{-5}
Ankle (Extragalactic Transition)	~1 EeV	Spectrum flattens	Transition point suggesting extragalactic dominance, with contributions from sources such as active galactic nuclei and gamma-ray bursts	$\sim 10^{-5}$–10^{-6}
Ultrahigh	~1 EeV–100 EeV	Likely extragalactic origins	Highest-energy cosmic rays that travel vast distances from extragalactic sources and potentially probe intergalactic magnetic fields	$\sim 10^{-6}$–10^{-8}

changes in the slope of the spectrum. These features are significant because they mark shifts in the underlying processes and potential sources of CRs at different energies. The following briefly describes the characteristics of CR spectra.

The first "knee" occurs around 1–3 PeV in energy and is characterized by a steepening of the spectrum. This is commonly interpreted as the energy limit of galactic sources, such as SNRs, which may no longer be capable of confining or accelerating particles above this threshold. Above the knee, fewer CRs are observed, suggesting either a shift in sources or limitations in galactic acceleration mechanisms.

The second "knee" (at ~100 PeV), sometimes called the "iron knee," appears at around 100 PeV and may correspond to the maximum energy of heavy nuclei, such as iron, within the galaxy. This feature suggests that different elements have different acceleration and confinement limits. The ankle (at ~1 EeV) marks a flattening of the spectrum after the steep decline above the knees. This transition signals the onset of extragalactic CRs, particles originating from sources beyond the Milky Way, possibly AGNs or gamma-ray bursts (GRBs). The ankle suggests that CRs from galactic sources are replaced or supplemented by higher-energy extragalactic sources.

Anomalies in CR flux at these different energy thresholds provide essential clues for fundamental physics, particularly for areas that remain open and unresolved, such as the so-called "dark matter" components of the cosmos whose existence or composition could be derived from the excess of light particles (e.g., electrons or positrons) in the CR spectra due to the interaction of the dark matter itself with the CRs of the ISM. Another similar investigation regards the so-called "primordial antimatter" by detection of anti-nuclei in CRs (e.g., anti-helium particles) whose presence would indicate the existence of primordial antimatter and potentially change our understanding of matter–antimatter asymmetry in the Universe.

Another component of SR whose origin is the galaxy is the so-called anomalous cosmic rays (ACRs), usually defined as those particles in the energy spectra of CRs that originate as interstellar neutral gas flowing into the heliosphere, becoming ionized, and eventually accelerated at the solar wind termination shock. As a result, the ACRs at low energies are mostly singly CPs that can also reach the surfaces of planets in our solar system [2].

An effective and widely used approach to characterize the CR flux in the heliosphere, specifically in deep space, where atmospheric interference is absent, is through the application of an empirical formula. This formula, which captures the energy dependence of CR, serves as a valuable tool in CR research. The differential flux, $\Phi(E)$, representing the number of particles per unit time, area, solid angle, and energy interval, is typically expressed by a power-law relationship:

$$\Phi(E) = \Phi_0 E^{-\gamma}$$

where E is the particle's energy, Φ_0 is a normalization constant (reflecting observed intensity at reference energy), and γ is the spectral index, usually between 2.6 and 3.0 for GCRs.

In deep space, we can calculate the total CR flux by considering the contributions of each particle species (protons, helium nuclei, etc.) separately and summing them. The total flux Φ_{tot} thus becomes

$$\Phi_{tot}(E) = \sum_i \Phi_{0,i} E^{-\gamma i}$$

where each particle type i has specific values for $\Phi_{0,i}$, and γ_i.

In this region of the heliosphere, CRs do not experience atmospheric attenuation, so the flux is independent of incidence angle and remains primarily isotropic, except for variations due to solar modulation. Further, in such a formula for CR flux, the knee and ankle represent changes in the spectral index γ at specific energy thresholds. The knee refers to a point in the energy spectrum where the power-law slope changes, and the ankle signifies a higher-energy threshold at which the slope changes again. These features are modeled by adjusting the power law slope γ in piecewise sections to reflect the observed changes in the spectrum.

GCRs are particularly interesting in space radiobiology, as they are continuously present in BLEO environments and can pass almost unimpeded through typical spacecraft shielding or the skin of an astronaut. This component of SR is a primary concern for BLEO missions requiring long travel durations, as GCRs, dominated by protons, pose significant exposure risks. However, the composition of GCRs can vary substantially. A critical question to be explored in a later chapter is how the energy range of GCRs relates to their potential biological impact on astronauts. Understanding which energies pose the most significant risks is of utmost importance. This analysis will be correlated with the specific exposure scenarios anticipated in deep-space missions, providing insight into the most hazardous aspects of cosmic radiation in BLEO environments.

2.2.2 *Solar radiation*

The Sun plays a crucial role in the low Earth orbit (LEO) and BLEO radiation environments, as some radiation components are generated directly by solar activity during SPEs [3, 4]. These events can produce high-intensity radiation emissions that are challenging to predict. Additionally, the Sun's magnetic field, which is a result of the Sun's internal dynamo process, influences CRs originating from our galaxy. This modulation of CR flux creates a long-term SR environment. The Sun's magnetic field, which varies in strength and polarity over an 11-year cycle, is tightly linked to the solar magnetic activity cycle, commonly known as the sunspot cycle. Superimposed on this 11-year cycle is a shorter-term periodicity known as the Bartels rotation (named after Julius Bartels, a German geophysicist and solar physicist) modulation. This arises from the Sun's roughly 27-day rotational period, during which active regions (ARs), such as sunspots and coronal holes, repeatedly face Earth. These regions influence the solar wind and the interplanetary magnetic field (IMF), creating periodic fluctuations in the flux of GCRs observed near Earth. Each Bartels rotation is assigned a sequential number for tracking solar phenomena over time.

The interaction of these periodic modulations with the broader solar cycle produces a complex pattern in space weather conditions. During high solar activity (solar maxima) periods, enhanced solar wind and stronger magnetic fields reduce CR penetration into the heliosphere, a phenomenon known as solar modulation. Conversely, when the Sun's magnetic field is weaker during solar minima, CR flux at Earth is higher. The 27-day Bartels rotation periodicity becomes particularly evident during these quieter phases, as recurring coronal structures dominate the modulation of CRs. Phenomena closely related to Sun-generated radiation include SFs, coronal mass ejections (CMEs), and solar energetic particles (SEPs).

SFs are powerful explosive events, characterized by sudden bursts of electromagnetic radiation and the rapid release of high-energy particles. SFs can significantly influence the ionosphere and LEO region, affecting radio communications. Additionally, these events often trigger reactions that may include CMEs and SEPs.

CMEs are substantial expulsions of plasma and magnetic fields from the Sun's corona, ejecting billions of tons of solar material and carrying a strong magnetic field that can exceed the background strength of the IMF.

CMEs travel outward from the Sun at speeds ranging from 250 km/s to nearly 3,000 km/s, with the fastest Earth-directed CMEs reaching our planet in 15–18 h, while slower CMEs may take several days. As they propagate away from the Sun, CMEs expand, and larger ones can reach nearly a quarter of the distance between the Earth and the Sun by the time they arrive.

SEPs are highly energetic CPs primarily composed of protons and some heavier ions and electrons, accelerated to significant speeds during intense solar events such as SFs and CMEs. SEPs can reach energies of several GeV and travel to Earth within minutes to hours after a solar event, making them one of the most immediate hazards for space missions. Unlike GCRs, SEPs are often more sporadic and are concentrated during solar storms, making them difficult to predict precisely. SEPs can penetrate spacecraft shielding, exposing astronauts to potentially harmful radiation doses, especially in BLEO missions when the Earth's magnetosphere offers less protection. These phenomena and their variability are illustrated in Figs. 2.1–2.3.

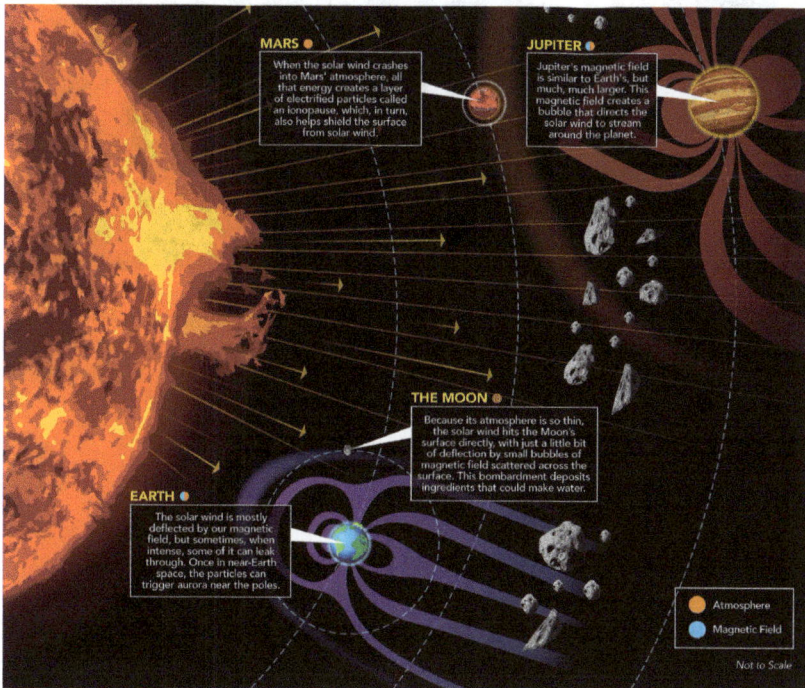

Fig. 2.1. Infographic showing solar wind across our solar system. Solar energetic particles are accelerated particles originating from the Sun during intense solar events such as flares and coronal mass ejections.

Source: NASA.

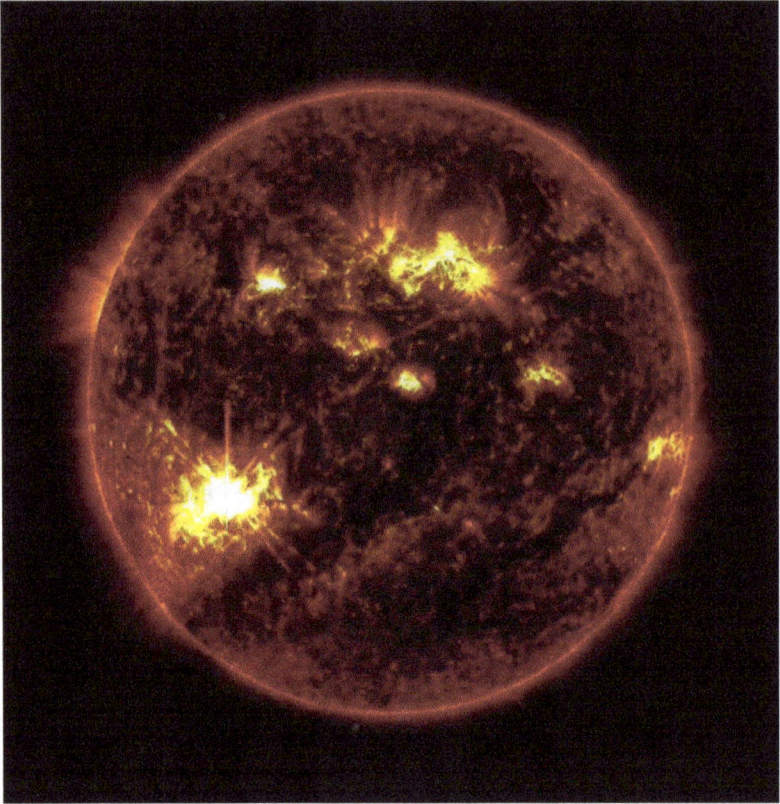

Fig. 2.2.   NASA's Solar Dynamics Observatory captured this image of a solar flare, as seen in the bright flash on the left, on 31 May 2024. The image shows a subset of extreme ultraviolet light that highlights the extremely hot material in flares and which is colorized in red.

Source: NASA/SDO.

This combination of solar activity-related events contributes to a dynamic, and sometimes hazardous, radiation environment in space. Understanding and predicting these phenomena remain crucial for ensuring the safety of both human-crewed and uncrewed space missions. The current approach to predicting SFs, usually starting events of or correlated to CMEs and SEPs, largely depends on monitoring the Sun's photospheric magnetic field in ARs. Most predictive models utilize parameters

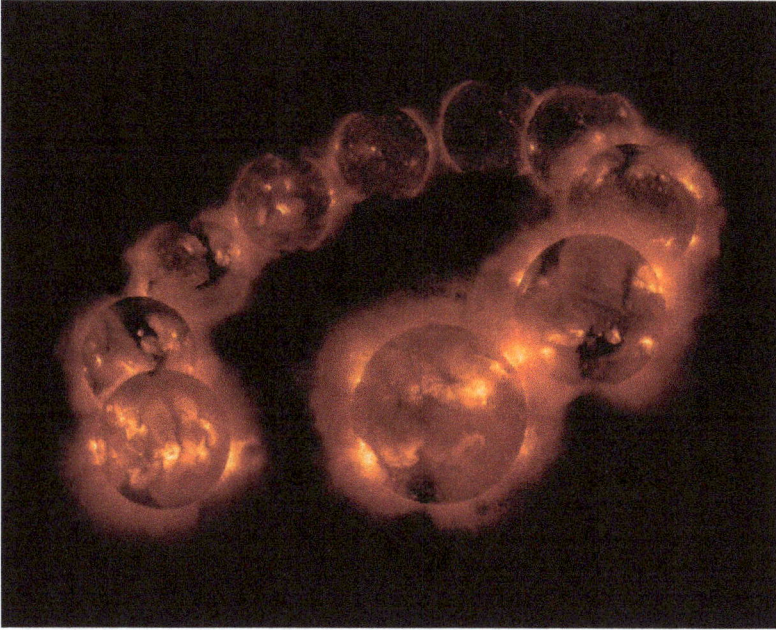

Fig. 2.3. A solar cycle: a montage of 10 years' worth of Yohkoh SXT images, demonstrating the variation in solar activity during a solar cycle, from 30 August 1991 to 6 September 2001.

Source: The Yohkoh mission of ISAS (Japan) and NASA (US). CC0 1.0 Universal (https://creativecommons.org/publicdomain/zero/1.0/deed.en).

extracted from existing datasets, such as the Space-Weather HMI Activity Region Patches (SHARP) [5], to estimate the probability of a flare. SHARP parameters provide comprehensive magnetic field data across entire ARs. However, they can sometimes overlook the nuanced evolution within the core areas of these regions, where large flares are likely to initiate. Researchers recognize this limitation and focus on more localized magnetic structures within ARs to improve flare predictions. The magnetic core, or the high-energy-density (HED) region, within an AR is believed to be critical in initiating large SFs. Researchers have developed new datasets targeting these HED regions by concentrating on areas with high free magnetic energy density in the photosphere, essentially the Sun's visible surface. This finer focus on high-energy zones improves the predictive capability of models by capturing changes in magnetic energy that may precede significant SFs [6]. This evolving approach to SF prediction demonstrates significant advances; it offers improved accuracy in forecasting the likelihood of SFs within a 24-hour window. However, while these models enhance our understanding of flare precursors, SF forecasting remains a complex task with room for further refinement. Challenges include the limited resolution of magnetic field measurements and the need for continuous model training to account for the Sun's dynamic behavior. As SF prediction capabilities improve, they could play an instrumental role in space weather forecasting, helping to mitigate the potential hazards posed to satellites, space missions, and terrestrial infrastructure.

2.2.3   *Trapped radiation (Van Allen belts)*

The Earth's magnetosphere acts as a protective shield, deflecting solar storms, the constantly streaming solar wind, and GCRs, which can harm technology and life on Earth. In deflection, the magnetosphere traps high-energy particles, which accumulate in two central regions of intense radiation: the inner Van Allen belt (IVAB) and the outer Van Allen belt (OVAB) – together constituting the Van Allen belts. These belts, primarily composed of relativistic electrons and protons, encircle the Earth like enormous doughnuts [7].

Astrophysicist James Van Allen discovered these belts in the late 1950s using instruments on board the Explorer 1 and Explorer 3 satellites, the first successful US space missions. Van Allen's team equipped these satellites with Geiger–Müller tubes, a type of particle detector used to measure the flux of CPs in space. As Explorer 1 entered the belts, the Geiger counters recorded unexpectedly high radiation levels, indicating the presence of trapped high-energy particles [8]. This groundbreaking discovery confirmed the existence of two distinct radiation belts, later named in honor of Van Allen.

Table 2.4 reports the VABs' primary origin, location, and composition characteristics.

Their intense radiation can cause satellite anomalies in space through deep-dielectric charging. The OVAB comprises billions of high-energy particles, which originate from the Sun; the IVAB results from interactions of CRs with Earth's atmosphere. The IVAB is produced by a combination of CR albedo neutron decay, which occurs when CRs scatter off the

Table 2.4.   Main characteristics of the Van Allen belts.

Feature	Inner Van Allen Belt	Outer Van Allen Belt
Origin	Trapping of high-energy particles within Earth's magnetic field	Interaction of solar wind particles with Earth's magnetic field
Distance from the Earth's Surface (km)	600–6,000	13,000–60,000
Composition	Primarily protons	Primarily electrons
Peak Radiation Levels	High proton flux (up to tens of MeV)	High electron flux (up to several MeV)
Dominant Source	Cosmic rays interacting with Earth's atmosphere	Solar wind and cosmic ray interactions with Earth's magnetosphere
Hazard to Spacecraft	Moderate to high, depending on shielding	High, due to high-energy electron flux

neutral atmosphere, producing neutrons that decay with a 15-minute half-life. Solar energetic protons, associated with flares and CMEs, also become trapped within the Earth's magnetic field. These energetic solar protons are the primary low-energy source (<50 MeV) for the IVAB and can sometimes produce a long-lived proton belt distinct from the inner zone. Astronauts traveling to the Moon on Apollo missions had to pass through the VABs, regions of intense radiation of trapped high-energy particles. Without careful planning, exposure in these belts could reach tens of millisieverts, even with protective aluminum shielding [9]. As an example of mitigating risk during the Apollo missions in the 1970s, NASA employed a combination of techniques that focused on optimizing flight trajectory through the VABs. In that sense, mission planners designed the flight trajectory to minimize the astronauts' time within the VABs. By selecting a path through the thinnest parts of these belts, where radiation density is lower, they reduced exposure to the trapped protons and electrons. This direct trajectory minimized travel time through the belts to only a few hours, significantly lowering the total radiation dose received. Also, they leveraged Earth's magnetosphere to benefit astronauts from its protective influence during certain parts of the mission. Though the Apollo missions ultimately traveled beyond the magnetosphere's shielding region, its residual protection substantially reduced CR exposure. This was beneficial when combined with other timing strategies.

There is a particular region of the IVAB known as the South Atlantic Anomaly (SAA), which has significant relevance for human space activities. This area originates from a misalignment of the Earth's rotational axis with that of its magnetosphere, resulting in an asymmetry of the magnetic field relative to the Earth's surface. This misalignment leads to a notably weaker magnetic field above the South Atlantic region, making it a unique and potentially hazardous area for space missions. The weakened magnetic field in the SAA allows CPs, primarily high-energy protons and electrons, to penetrate deeper into Earth's atmosphere compared to other regions. As a result, the SAA experiences a higher flux of these CPs, creating a more intense radiation environment than surrounding areas. This phenomenon can significantly increase radiation exposure for spacecraft and astronauts flying at altitudes between approximately 200 and 1,000 km, where LEO space missions, including the International Space Station (ISS), typically operate. As the ISS orbits the Earth, it regularly flies through the SAA, which exposes its crew to elevated radiation levels during these passages. Due to its geographical location and the resulting magnetic field anomalies, the SAA poses unique challenges for satellite operations and crewed missions. Satellites passing through this region may experience increased radiation levels, leading to malfunctions in onboard electronics, increased background noise in scientific instruments, and potential risks to astronaut health. To mitigate these risks, mission planners must consider flight paths that minimize time spent in the SAA. Additionally, spacecraft operating in this region are often equipped with enhanced shielding and radiation monitoring systems to protect the crew and sensitive equipment from elevated radiation exposure.

2.3   Space Radioprotection

Radioprotection is a field dedicated to minimizing radiation exposure to protect human health, often guided by the ALARA principle: as low as reasonably achievable. This principle is rooted in three main strategies for reducing radiation exposure:

- *Increasing distance from radiation sources*: Exposure intensity decreases with distance from the source, so increasing distance is an effective means of protection.
- *Limiting exposure time*: Reducing the duration of exposure helps minimize the total radiation dose.

- *Creating shielding*: Appropriate shielding materials block or absorb radiation, protecting individuals in the surrounding environment.

These strategies are widely used on Earth in environments where radiation is present, such as radiotherapy rooms, nuclear medicine applications, and nuclear power plants.

In space exploration, applying these radioprotection principles is much more challenging due to several unique factors:

- *Distance*: Unlike in controlled environments on Earth, space travelers cannot quickly increase their distance from radiation sources, as cosmic radiation and SPEs are pervasive and omnipresent.
- *Exposure time*: Space missions are characterized by long-term exposure to SR, particularly those involving deep-space travel or extended stays on the Moon or Mars. Limiting time in these environments is not feasible with current space exploration timelines.
- *Shielding limitations*: SR, primarily composed of high-energy CRs and solar particles, is more intense than most Earth-based radiation sources and comprises a broad range of particle types and energy levels (non-monoisotopic and non-monoenergetic). Effective shielding against these high-energy particles requires heavy, dense materials, which are often impractical to transport and use in spacecraft or habitats due to weight constraints.

These factors make it difficult to fully implement ALARA-based strategies in space exploration. Therefore, development of alternative, innovative radioprotection techniques specifically adapted to space conditions is required. Table 2.5 compares the effectiveness and feasibility of the three primary radioprotection principles – increasing distance, limiting exposure time, and creating shielding – across different radiation exposure scenarios: radiotherapy rooms, nuclear medicine applications, nuclear plants, deep-space travel, and living on a Moon or Mars base. Each column represents one of these core strategies and illustrates how practical or challenging each is in the specified environments.

In summary, while Earth-based radioprotection strategies work well in controlled environments, they become limited or impractical for space exploration. The table highlights the need for new, innovative radioprotection approaches specifically designed to tackle the challenges of deep space and extraterrestrial environments, where conventional methods cannot fully address SR's pervasive and complex nature. In the

Table 2.5.   Comparison of radioprotection strategies across different environments.

Environment	Increase Distance	Limit Exposure Time	Create Shielding
Radiotherapy Rooms	Feasible. Room design enables a safe distance from radiation sources.	Feasible. Managed exposure times for staff and patients.	Feasible. Lead-lined walls provide adequate protection.
Nuclear Medicine	Feasible. Targeted source locations allow a safe distance.	Feasible. Carefully monitored exposure durations for safety.	Feasible. Shielding targets from specific radiation sources.
Nuclear Plants	Feasible. Facilities kept in non-essential areas far from sources.	Feasible. Staff exposure is monitored and limited by regulations.	Feasible. Thick concrete/lead blocks radiation effectively.
Deep-Space Travel	Not feasible. Space radiation is omnipresent. Distance offers no protection.	Not feasible. Prolonged exposure is inherent to deep-space missions.	Partially feasible. Weight limits restrict shielding; partial shielding helps only against solar events.
Moon/Mars Bases	Not feasible. Cosmic radiation permeates the habitat regardless of location.	Partially feasible. Work shifts may limit exposure, but long-term exposure is unavoidable.	Partially feasible. Underground habitats or advanced materials may reduce exposure, but complete protection is challenging.

following, the unique radioprotection challenges on the Moon's surface, Mars' surface, and in free space, as well as the considerations for astronaut suits, starships, and space stations, are discussed.

2.3.1 *Lunar surface environment*

The lunar surface presents a high-radiation environment due to the lack of a protective atmosphere or significant magnetic field. Unlike Earth's thick atmosphere and strong dipolar magnetic field, which together act as a shield against CPs, the Moon has only an extremely thin exosphere composed mainly of hydrogen, neon, and argon. CRs and SEPs reach the lunar surface directly, producing secondary particles upon impact with the regolith. These secondary particles include neutrons, gamma rays, and other subatomic particles that can penetrate habitats or astronaut suits, potentially increasing radiation exposure.

Hence, the CP radiation environment on the lunar surface consists of GCRs, a small contribution from ACRs, and a highly variable, sporadic contribution from SEPs. In addition, secondary albedo particles are created primarily by the interaction of GCRs with the lunar regolith, including neutrons and protons. Some particles can escape from the soil and be measured as albedo particles. They also contribute to astronauts' radiation exposure on the lunar surface [10, 11].

Radioprotection strategies on the Moon emphasize shielding structures using *in situ* materials such as regolith, which could help absorb primary and secondary particles, thereby reducing exposure to harmful radiation levels. Lunar regolith has been characterized for its potential capability to effectively shield in the context of lunar lava tubes [12]. Lunar lava tubes are naturally occurring tunnels beneath the Moon's surface, believed to have formed billions of years ago during ancient volcanic activity. When molten lava flowed across the Moon's surface, the outer layers cooled and solidified, while the interior continued to flow. Eventually, when the lava flow ceased, it left behind hollow, tube-like structures that formed these lava tubes. Lunar lava tubes can vary widely, with some theorized to be large enough to house entire habitats, providing a naturally sheltered environment that could offer significant protection from SR. Lava tubes, with their solid rock ceilings and natural insulation, could serve as protective shelters, significantly reducing radiation exposure by absorbing primary particles and mitigating secondary particle production. Lunar regolith – the delicate, powdery dust covering the Moon's surface – has unique properties that make it suitable for shielding against harmful radiation. Its composition includes silicate minerals and small amounts of metallic iron, which can effectively attenuate radiation. Using lunar regolith as an additional shielding layer within or over lava tube habitats could help absorb incoming particles, thereby providing even better protection for lunar explorers. Studies have shown that even modest regolith layers can reduce radiation levels by as much as 50% or more, especially in combination with the solid rock of lava tube structures.

Several space missions and projects are underway or in development to identify, map, and characterize lunar lava tubes as potential sites for future habitats:

- *SELENE (Kaguya)*: Japan's SELENE mission, launched in 2007, was one of the first to identify the existence of sizeable potential lava tubes on the Moon using radar technology. The mission's observations provided initial insights into these tubes' structural stability and depth.
- *Lunar Reconnaissance Orbiter (LRO)*: NASA's LRO, launched in 2009, has extensively mapped the lunar surface and identified possible entrances to lava tubes using high-resolution imagery and radar data. These data have been critical for understanding the size and distribution of lunar lava tubes and assessing their potential as future shelters.
- *Artemis Program*: NASA's Artemis program, which aims to return humans to the Moon, is planning missions involving detailed exploration of the Moon's surface. As part of this effort, potential sites for sustainable human habitation, including lava tubes, are being studied. Artemis missions are expected to provide ground-based assessments of these tubes in the coming years.

2.3.2 *Mars surface environment*

Mars presents a similarly high-radiation environment to the Moon, though with a few unique characteristics. The planet has a thin carbon dioxide-based atmosphere that offers minimal protection against solar and cosmic radiation – roughly 1% of the Earth's atmospheric volume. While the Martian atmosphere does absorb some incoming particles, it is not dense enough to shield against high-energy CRs and SEPs. When these particles interact with the Martian atmosphere, they produce secondary radiation, creating a complex environment on the surface that astronauts must contend with.

In addition, Mars lacks a global magnetic field, limiting natural protection against cosmic radiation, just as on the Moon. The absence of a magnetic field and a dense atmosphere means that solar radiation on the Martian surface is significantly higher than on Earth. Current estimations forecast that exposure to SR on Mars' surface may be 10^2–10^3 times higher than on Earth's surface [13, 14].

One potential strategy for radiation protection on Mars is using Martian regolith, the loose, fragmented surface material composed of dust, sand, and small rocks, as shielding. Martian regolith contains various minerals, including silicates, oxides, and potentially tiny amounts of water in the form of hydrated minerals. Studies on Martian analog materials suggest that regolith layers of sufficient thickness could reduce radiation levels inside habitats. Covering habitats with regolith or using it as a construction material may reduce the dose of CRs and SEPs that penetrate living spaces, thereby providing substantial radiation protection for astronauts. Unlike Earth, Martian regolith is also believed to contain certain materials that might be particularly useful for building sturdy, protective structures, possibly through sintering or 3D printing technologies. Using regolith as a primary source for building materials would significantly reduce the need to transport heavy shielding materials from Earth – a significant logistical advantage for sustained missions.

Mars also has extensive lava tube networks that could offer natural protection from radiation. Similar to those on the Moon, these lava tubes are believed to have formed from volcanic activity billions of years ago. As lava flowed across the Martian surface, the outer layers cooled and solidified, while the molten interior continued to flow, eventually leaving behind hollow tunnels. Recent radar and imaging data from Mars orbiters suggest the presence of large lava tube networks on Mars, with some tubes potentially wide and tall enough to house habitats or equipment.

The shielding provided by lava tubes could significantly reduce radiation exposure for astronauts by providing a natural barrier of rock that absorbs incoming CRs and SEPs. Studies have shown that even a modest thickness of rock overhead can reduce radiation levels, and the solid rock structure of lava tubes would add stability and thermal insulation, both of which are valuable for a Martian habitat. Several current and planned Mars missions are focused on understanding and mapping the Martian subsurface, including lava tubes, to assess their potential as shelter sites. Among the more relevant are listed as follows:

- *Mars Reconnaissance Orbiter (MRO)* [15]: NASA's MRO has been crucial in mapping potential lava tubes and cave entrances on Mars using high-resolution imaging and radar data. These images provide insights into surface formations that could lead to subsurface tunnels.

- *ExoMars Program* [16]: An original joint mission between ESA and Roscosmos, ExoMars is designed to search for signs of life on Mars and study its subsurface properties. The Trace Gas Orbiter (TGO) and future lander missions aim to investigate subsurface structures, including caves and lava tubes, as potential microbial habitats or future human shelters.
- *Mars Ice Mapper* [17]: This proposed mission aims to map water ice on Mars, including subsurface reservoirs within lava tubes or caves. The presence of ice could provide valuable resources for habitat support, potentially reducing the need to transport large amounts of water from Earth.
- *Mars Sample Return Mission* [18]: This mission, focused on collecting samples from Mars and returning them to Earth, will help characterize the Martian surface and subsurface environment, aiding future habitat planning.

By combining *in situ* resources such as Martian regolith and identifying natural shelters such as lava tubes, future missions can develop a more sustainable approach to radiation protection on Mars. This approach could enable safer, longer-term human exploration.

2.3.3 *Lunar vs. Martian regolith*

Table 2.6 highlights the different lunar and Martian regolith compositions, which also influence the capability of creating shielding or protective infrastructure using construction materials containing such components. The distinct elemental compositions of lunar and Martian regoliths impact their potential effectiveness in radiation protection for human habitats. Here are some key considerations based on the differences:

- *Iron content*: Martian regolith has a higher iron content (around 19%) compared to lunar regolith (approximately 13%), which could make it more effective in absorbing certain types of ionizing radiation, such as GCRs. Iron-rich materials can contribute to shielding but may also generate secondary radiation. Thus, regolith with high iron may offer some radiation protection but also require careful design to mitigate secondary particle risks.

Table 2.6. Comparison of the elemental composition of lunar and Martian regoliths.

Element	Lunar Regolith (wt %)	Martian Regolith (wt %)
Oxygen (O)	~41–45	~43
Silicon (Si)	~20–21	~20–21
Aluminum (Al)	~5,~10–13 (site variable)	~3–5
Calcium (Ca)	~8–10	~3–4
Iron (Fe)	~6/~13/~15 (site variable)	~12–15
Magnesium (Mg)	~5	~5
Titanium (Ti)	~1–5 (site variable)	~0.5–07
Sodium (Na)	~0.6	~3
Potassium (K)	Trace (<0.5)	~0.6
Sulfur (S)	Trace	~2–3
Carbon (C)	Trace	Not detected
Chlorine (Cl)	Trace	~0.6–0.7

Sources: Lunar Sourcebook (LPI, 1991); NASA Apollo mission sample analyses; Mars Pathfinder &
Viking lander soil composition data (NASA/JPL); Curiosity rover APXS measurements; USGS &
LPI regolith composition datasets.

- *Sulfur content*: Martian regolith contains around 3% sulfur, much higher than lunar regolith. This sulfur could be used to create sulfur-based concrete for building Martian structures, providing physical and radiation protection. On the Moon, sulfur content is minimal, so other binding materials or techniques would be needed to create similar structures.
- *Oxygen and silicon compounds*: Both regoliths are rich in oxygen and silicon, with oxygen found mostly within silicate minerals (e.g., SiO_2 compounds). These silicates contribute to overall structural integrity and provide moderate radiation shielding due to their density and ability to absorb secondary particles. However, the effectiveness of these materials in blocking highly energetic CRs remains limited, so additional shielding layers will be necessary.
- *Titanium in lunar regolith*: Lunar regolith, especially in the basaltic plains known as "lunar maria," often contains more titanium, which has good radiation-blocking properties. Titanium-rich minerals, such as ilmenite, can offer increased radiation protection compared to the silicate-dominated regolith on Mars. This titanium presence on the Moon could aid in developing effective shielding materials, either by processing ilmenite directly or by using it to bind regolith layers.

Hence, both lunar and Martian regoliths can be used as raw materials for creating artificial radiation shields, with Mars' sulfur content supporting sulfur-based concrete. At the same time, the Moon's titanium and ilmenite are more suited to sintering or other mineral-processing methods. Studies suggest that a regolith layer of around 2–3 m could significantly reduce exposure to harmful CRs, thereby providing long-term protection for human habitats.

2.3.4 *Free space*

In deep space, away from any planetary surface, astronauts are continuously exposed to GCRs and potentially intense SPEs, especially during the solar maximum. The lack of any atmospheric or magnetic shielding means that radiation exposure is highest in free space, posing risks of acute radiation sickness and long-term health effects. Secondary radiation in this context can arise from the interaction of high-energy particles with spacecraft materials, producing neutrons and gamma rays within the cabin. Radioprotection strategies in free space require robust shielding, radiation forecasting, and potential "storm shelters" within spacecraft for protection during intense solar events. It is possible to distinguish two areas of research and technological developments that will shape the future astronaut suits, starships, and space stations.

Space suits designed for extravehicular activities (EVAs) must provide sufficient radioprotection to safeguard astronauts working in environments exposed to high radiation levels, whether on the lunar surface, on Mars, or during spacewalks in free space. Current EVA suits are layered to offer some radiation protection, but improving these suits to shield against both primary and secondary radiation remains a major engineering challenge. Advanced suit designs explore materials that can better absorb or deflect CRs and secondary radiation without adding prohibitive weight and incorporate sensors to monitor radiation dose in real time. Space suits incorporate radiation shielding materials strategically placed in areas where the body is most vulnerable. These materials possess high atomic numbers and densities, which allow them to absorb and scatter radiation particles, thereby reducing their penetration into the suit and the astronaut's body. Commonly used radiation shielding materials in space suits include high-density polyethylene (HDPE), aluminum, and composite materials. HDPE is particularly effective against high-energy protons and provides good shielding properties while maintaining its lightweight characteristic. Aluminum is another commonly used material due to its high atomic number and effectiveness in shielding against gamma rays and some high-energy particles. It is often used in the helmet visor and in other critical areas. Composite materials, made by combining different materials, can balance weight and shielding effectiveness. These composites can incorporate layers of metals, such as aluminum or titanium, with polymer-based materials to provide efficient radiation protection while maintaining a manageable weight. The thickness and placement of the shielding layers in space suits are determined through extensive modeling and simulation, considering the specific radiation environment and mission parameters. This ensures that the suits provide adequate protection without compromising mobility or adding excessive weight. It's important to note that eliminating radiation exposure is not feasible, but the goal is to minimize the dose and associated risks to acceptable levels. Continuous research and technological advancements in radiation shielding materials and designs enhance astronaut safety during space missions, especially for BLEO human exploration [19].

Spacecraft and space stations are essential for long-duration missions, providing a controlled environment with layered shielding. Structures such as the ISS are situated in LEO, where the Earth's magnetosphere offers some protection against GCRs and SPEs. However, spacecraft will face unmitigated cosmic and solar radiation in BLEO missions, such as those to Mars. Spacecraft materials can interact with incoming radiation, creating secondary particles within the habitat. Effective radioprotection in starships and future space stations will include reinforced hulls with advanced materials, radiation "storm shelters," and potentially hydrogen-rich materials to shield against primary and secondary radiation. These measures are critical to reducing astronauts' cumulative dose on prolonged missions. SR protection of complex structures [20] such as starships, space stations, or, in the future, the first settlements on the Moon or Mars is a complex task that can be addressed using a passive shielding protection approach with the same materials described previously.

Radiation exposure significantly affects metals and alloys. It primarily leads to the formation of lattice vacancies and interstitial atoms within the crystal structure, resulting in increased yield strength and decreased fracture toughness. The elastic properties of metals, such as their moduli, remain relatively unaffected by fast neutrons unless the fluence exceeds

10^{17} n/cm². However, radiation markedly influences plastic properties, reducing their plasticity and flexibility while increasing hardness.

These effects are attributed to the hindrance of dislocation motion, as plasticity is closely associated with dislocation movement. Various factors influence the impact of neutron irradiation on mechanical properties, including exposure temperature, duration, fluence, energy spectrum, and material properties such as composition, cold work, heat treatment, and grain size.

Neutron irradiation mainly affects tensile and elongation properties, fatigue, hardness, reduction in area, and the transition from brittle to ductile fracture. The changes depend on the metal type and factors such as fluence, irradiation, test duration, and temperature. Creep rate and stress-rupture properties are also generally affected.

One significant consequence of radiation embrittlement in structural materials is reduced fracture toughness. Monitoring the degradation of fracture toughness through a surveillance program is crucial for ensuring the safe operation of spacecraft throughout their lifetimes. Furthermore, developing embrittlement prediction models is essential for estimating and extending the lifespan of spacecraft and establishing effective surveillance programs for aging spacecraft.

Recently [21], research in this field has been moving toward a new radiation shielding evaluation method that calibrates the model with actual radiation exposure data from the ISS to provide a more accurate assessment of radiation shielding requirements for operations in cis-lunar space and establishes an empirical link to the medical histories of the ISS astronauts to assess long-term radiation exposure hazards better. Such methods are used to evaluate the effectiveness of various materials and shielding designs for passive radiation shielding from GCRs for spacecraft and space habitats, and they are shown to significantly reduce the uncertainty in determining the mass of shielding material required to obtain acceptable exposure levels.

Also, inflatable spacecraft [22] are considered a cost-effective solution for human activities in space, offering advantages in radiation protection and cost reduction compared to traditional aluminum construction. An example is the TransHab design by NASA's Johnson Space Center. Its wall layup combines beta cloth, aluminized Kapton, aluminized Mylar, Nextel fabric, open-cell foam, Combitherm, and Nomex. The compositions and areal densities of two inflatable designs are compared to those of aluminum and HDPE shields to evaluate their effectiveness in reducing

SR exposures. HDPE is found to be one of the better polymers for radiation reduction. Only HDPE shows a reduction in dose equivalent at lower thicknesses compared to free-space exposure, while at larger thicknesses, all three materials exhibit some reduction. HDPE performs better in reducing exposure risk, based on tumor prevalence (TP), cell death, and transformation models. The TransHab inflatable layup performs slightly better than aluminum as a shielding material. Still, pure HDPE performs significantly better due to its higher hydrogen content. Further analysis is needed to explore improved polymers for spacecraft construction, and the evaluation of materials for GCR must include biological response models. To provide an optimistic dose equivalent estimate for future Mars missions, pure HDPE is selected as a high-performing shielding material that may closely resemble future designs.

Distributed measurements are crucial in cis-lunar space to understand energetic particles. This includes measurements in both the inner and outer heliosphere and magnetosphere to capture key physical elements and spatial scales. Improving the understanding and forecasting of SEP events requires simultaneously imaging more of the solar surface and sampling particles and magnetic fields. Consistent measurements of the radiation environment within the magnetosphere are necessary. Daily average fluxes derived from different instruments can vary significantly, so continued measurements are needed to understand and forecast radiation belt fluxes, especially as Solar Cycle 25 intensifies. The Gateway, stationed in cis-lunar space, will experience higher GCR intensities and more extreme particle intensity and dose variations than the ISS. Co-located measurements of particle transport through materials are essential for designing future space and lunar missions. The Gateway's instrumentation provides simultaneous flux and dose measurements, contributing to a comprehensive understanding of the unique environment. Interactions between energetic particles and the regolith on the lunar surface create albedo particles. Previous measurements of lunar albedo neutrons covered energies only up to ~20 MeV; therefore, future experiments should extend these measurements to higher energies to fully characterize the lunar surface environment [23, 24].

2.3.5 *Active shielding: emulating nature's "approach"*

To sustain life on Earth, nature has provided two fundamental mechanisms for shielding against harmful cosmic radiation: passive and active shielding. These natural defenses work differently but, together, create a habitable environment that has allowed life to flourish and evolve safely over millions of years.

Like Earth's dense atmosphere, passive shielding absorbs and blocks radiation, preventing high-energy particles from reaching the surface. This shielding type acts as a physical barrier, naturally absorbing much of the dangerous radiation from space. Passive shielding is effective because it uses mass, such as the thick layers of atmospheric gases, to diminish radiation intensity gradually. This barrier approach has allowed Earth's surface to remain relatively safe from CRs and solar radiation, thereby creating a stable environment for life to evolve and thrive.

In contrast, active shielding is represented by Earth's magnetic field, which dynamically deflects CPs away from the planet. The magnetic field creates a protective zone called the magnetosphere, where CPs are redirected along magnetic lines, keeping them from penetrating the atmosphere deeply. This active shielding is well suited for defending against CPs because it works like an electromagnetic force field, repelling them through redirection rather than absorbing them. The magnetosphere's dynamic nature allows it to continuously adapt to changes in space weather, including solar storms, providing an additional layer of protection for life on Earth. This dual-layered system of passive and active shielding has been instrumental in human survival, allowing the human species to develop without harmful genetic or cellular damage levels. As humanity begins exploring long-duration space missions to Mars or beyond, replicating these natural shielding mechanisms becomes essential. Scientists are now studying ways to mimic Earth's defenses through next-generation active shielding technologies that can protect humans from radiation in space, where the atmosphere and magnetic field are absent. These efforts aim to recreate Earth's natural protections, safeguarding astronauts from SR and making interplanetary exploration safer for human life.

So, next-generation space exploration will eventually require active shielding, a possible approach that has been under study and evaluation for many years with the aim of providing additional protection against SR by actively mitigating or deflecting radiation particles. Unlike passive shielding, which relies on materials to absorb or block radiation, active shielding involves the use of active systems and technologies to reduce radiation exposure. Here, we present some key aspects of active shielding that are being investigated.

One active shielding approach involves using magnetic fields to deflect CPs. By generating strong magnetic fields around the spacecraft or inhabited areas, CPs are deflected away from the protected region. This method is particularly effective against CPs, such as protons and electrons [25, 26]. A solenoidal active magnetic shielding design analysis was performed to determine its effectiveness in reducing the dose equivalent from GCRs. This analysis provides insights into the potential effectiveness of solenoidal active magnetic shielding for reducing radiation exposure during space exploration missions, including those involving starships [27].

Other possible approaches to active shielding are as follows:

- *Electrostatic shielding*: It involves using electric fields to repel CPs. By creating electric fields around the spacecraft or living spaces, CPs can be deflected or repelled, reducing their penetration into the protected area [28].
- *Active radiation monitoring*: Active shielding systems often incorporate advanced radiation monitoring and detection systems. These systems continuously monitor the radiation environment and provide real-time data on radiation levels. This information allows for proactive adjustments and countermeasures to minimize radiation exposure.
- *Particle beam deflection*: Some active shielding concepts propose using directed particle beams to counteract incoming radiation particles. By generating and directing high-energy particle beams toward incoming particles, they can be deflected or neutralized before reaching the protected area [29].
- *Plasma-based shielding*: Plasma-based shielding is another active shielding concept that involves creating a plasma sheath around the spacecraft or habitat. This plasma sheath interacts with incoming radiation, scattering and absorbing particles, thus providing additional protection [30].

Active shielding technologies are still in the early stages of development and face significant challenges, such as power requirements, scalability, and system complexity. However, they promise to enhance radiation protection during long-duration space missions, particularly for journeys to Mars or beyond, where radiation exposure is a significant concern.

Further research and development are needed to refine and validate active shielding concepts, ensuring their effectiveness, reliability, and practicality for space exploration.

References

[1] Rahmanifard, F., de Wet, W. C., Schwadron, N. A., Owens, M. J., Jordan, A. P., Wilson, J. K., *et al.* (2020). Galactic cosmic radiation in the interplanetary space through a modern secular minimum. *Space Weather*, 18(9), e2019SW002428. https://doi.org/10.1029/2019SW002428.

[2] Cummings, A. C., Stone, E. C., and Steenberg, C. D. (2002). Composition of anomalous cosmic rays and other heliospheric ions. *The Astrophysical Journal*, 578, 194–211.

[3] Malandraki, O. E. and Crosby, N. B. (2018). *Solar Particle Radiation Storms: Forecasting and Analysis. The HESPERIA HORIZON 2020 Project and Beyond.* Springer, Cham, Switzerland. https://doi.org/10.1007/978-3-319-60051-2.

[4] Gleeson, L. J. and Axford, W. I. (1968). Solar modulation of galactic cosmic rays. *Astrophysical Journal*, 154, 1011. https://doi.org/10.1086/149822.

[5] Bobra, M. G., Sun, X., Hoeksema, J. T., *et al.* (2014). The Helioseismic and Magnetic Imager (HMI) vector magnetic field pipeline: SHARPs – Space-Weather HMI Active Region Patches. *Solar Physics*, 289, 3549–3578. https://doi.org/10.1007/s11207-014-0529-3.

[6] Li, T., *et al.* (2024). Survey of magnetic field parameters associated with large solar flares. *The Astrophysical Journal*, 964(159), 7. https://doi.org/10.3847/1538-4357/ad2e90.

[7] Li, W. and Hudson, M. K. (2019). Earth's Van Allen radiation belts: From discovery to the Van Allen Probes era. https://doi.org/10.1029/2018JA025940.

[8] Baker, D. N. and Panasyuk, M. I. (2017). Discovering Earth's radiation belts. *Physics Today*, 70(12), 46–51. https://doi.org/10.1063/PT.3.3791.

[9] Marki, A. (2020). Radiation analysis for Moon and Mars missions. *International Journal of Astrophysics and Space Science*, 8(3), 16–26. https://doi.org/10.11648/j.ijass.20200803.11.

[10] Xu, Z., *et al.* (2022). Primary and albedo protons detected by the Lunar Lander Neutron and Dosimetry experiment on the lunar farside, *Front. Astron. Space Sci., Sec. Space Physics*, 9, 974946. https://doi.org/10.3389/fspas.2022.974946.

[11] Dobynde, M. I. and Guo, J. (2021). Radiation environment at the surface and subsurface of the Moon: Model development and validation. *Journal of Geophysical Research: Planets*, 126, e2021JE006930. https://doi.org/10.1029/2021JE006930.

[12] Feng, Y., *et al.* (2024). A comprehensive review of the construction of lunar lava tube base and field research on a potential Earth test site. *International Journal of Mining Science and Technology*, 34(9), 1201–1216. https://doi.org/10.1016/j.ijmst.2024.06.003.

[13] Baker, D. M., H., Wouter, de Wet, Bent, E., and *et al.* (2017). The radiation environment on the surface of Mars – Summary of model calculations and comparison to RAD data. *Life Sciences in Space Research*, 14, 18–28. https://doi.org/10.1016/j.lssr.2017.06.003.

[14] Knutsen, E. W., *et al.* (2021). Galactic cosmic ray modulation at Mars and beyond was measured with EDACs on Mars Express and Rosetta. *Astronomy & Astrophysics*, 650, A165. https://doi.org/10.1051/0004-6361/202140767.

[15] Jafarzade, R. (2024). Mars exploration – Science, instruments, and technologies. *International Astronautical Conference 2023 Conference Proceedings*.

[16] Drahl, C. (2023). The long-awaited mission that could transform our understanding of Mars. *Knowable Magazine|Annual Reviews*. https://doi.org/10.1146/knowable-050323-1.

[17] Amoroso, E., *et al.* (2024). International Mars Ice Mapper Mission: A step forward to map the subsurface water ice and prepare for future human Mars exploration. *International Astronautical Conference 2024 Conference Proceedings*.

[18] Loureiro, T. (2024). Mars Sample Return – Status of the Earth Return Orbiter Mission. *International Astronautical Conference 2024 Conference Proceedings*.

[19] Waterman, G. (2019). AstroRad radiation protective equipment evaluations on Orion and ISS. *Proceedings of the 70th International Astronautical Conference 2019 (IAC2019)*, Washington D.C., 21-25 October 2019, USA.

[20] Wang, J. J., Singleterry, R. C. Jr., Ellis, R. J., and Hunter, H. T. (2002). Radiation effects on spacecraft structural materials. *Proceedings of International Conference on Advanced Nuclear Power Plants (ICAPP)*, Embedded International Topical Meeting 2002 ANS Annual Meeting, June 9-13, 2002, Hollywood, Florida.

[21] Warden, D. (2019). *Analysis of Passive Radiation Shielding from Galactic Cosmic Rays in Cis-Lunar Space*. Rice University ProQuest Dissertations Publishing, Houston, Texas, 2019, 28735637.

[22] Valle, G., Litteken, D., and Jones, T. C. (2019). Review of habitable softgoods inflatable design, analysis, testing, and potential space applications. *Proceedings of the Science and Technology Forum and Exposition*, San Diego, January 2019.

[23] Corti, C., Whindam, K., Ravindra, D., Jamie, R., Du Toit, S., Nariaki, N., Drew, T., and Thomas, Y. C. (2022). Galactic cosmic rays and solar energetic particles in cis-lunar space: Need for contextual energetic particle measurements at Earth and supporting distributed observations. https://doi.org/10.48550/arXiv.2209.03635.

[24] Zaman, F. A. and Townsend, L. W. (2021). Radiation risks in cis-lunar space for a solar particle event similar to the February 1956 event. *Aerospace*, 8, 107. https://doi.org/10.3390/aerospace8040107.

[25] Shephard, S. G. and Kress, B. T. (2007). Störmer theory applied to magnetic spacecraft shielding. *Space Weather*, 5(4), S04001.

[26] Ferrone, K., Willis, C., Guan, F., Ma, J., Peterson, L., and Kry, S. (2023). A review of magnetic shielding technology for space radiation. *Radiation*, 3, 46–57. https://doi.org/10.3390/radiation3010005.

[27] Battiston, R., *et al.* (2012). Active radiation shield for space exploration missions. *Final Report ESTEC Contract N° 4200023087/10/NL/AF: "Superconductive Magnet for Radiation Shielding of Human Spacecraft."*

[28] Chowdhury, R. P., Stegeman, L., Santillana Padilla, R. F., Lund, M. L., Madzunkov, S., Fry, D., and Bahadori, A. A. (2021). Space radiation electrostatic shielding scaling laws: Beam-like and isotropic angular distributions. *Journal of Applied Physics*, 130, 034903. https://doi.org/10.1063/5.0046599.

[29] Washburn, S. A., Blattnig, S. R., Singleterry, R. C., and Westover, S. C. (2015). Active magnetic radiation shielding system analysis and key technologies. *Life Sciences in Space Research*, 22–34. https://doi.org/10.1016/j.lssr.2014.12.004.

[30] Adams, J. H., *et al.* (2005). Revolutionary concepts of radiation shielding for human exploration of space. U.S. National Aeronautics and Space Administration, NASA Technical Report NASA/TM-2005-213688, March 2005.

Chapter 3

Measuring Space Radiation

3.1 Introduction

The foundation of any scientific endeavor is the ability to measure natural phenomena with great accuracy and precision. This principle applies to ionizing radiation (IR), and in this section, we delve into the fundamentals of measuring space radiation (SR) since its inception. SR measurement is crucial for two primary purposes: understanding the physical quantities characterizing cosmic rays (CRs) and monitoring radiation to calculate dosimetry quantities for human safety. This section explores the methods and instruments used to measure and quantify astronauts' SR exposure, ensuring their safety during missions. First, we examine the measurements to investigate the physical properties of CRs, which are high-energy particles originating from outer space. These measurements focus on understanding CRs' origins, composition, and energy spectra. Instruments such as particle detectors, spectrometers, and telescopes are utilized in various space missions to capture and analyze CR data. These instruments provide critical insights into fundamental physics and cosmology, helping to address open questions about the Universe's structure and the processes occurring within it. The following chapter will discuss the measurements designed to monitor SR and calculate dosimetry quantities. These measurements are essential for assessing astronauts' radiation exposure during space missions. Dosimetry focuses on quantifying the absorbed dose, equivalent dose, and effective dose, which are key indicators of the potential biological effects of radiation on human tissues. Instruments such as dosimeters, tissue-equivalent proportional counters, and radiation area monitors provide real-time monitoring and accurate dose assessment, ensuring that astronauts remain within safe exposure limits. After reviewing the historical and future space missions dedicated to measuring SR, we highlight the potential complementarity of the instruments used in these missions with those discussed in the previous section. The interplay between scientific research instruments and dosimetry tools enhances our ability to advance our understanding of CRs and protect human health in space. By distinguishing between the two main objectives of SR measurement, advancing fundamental scientific knowledge and ensuring astronaut safety, we can appreciate the comprehensive approach required to tackle the challenges of space exploration. This dual focus underscores the importance of precision and innovation in developing and deploying the instruments that enable accurate measurement and understanding of SR.

3.2 Astroparticle Experiments in Space

Galactic cosmic rays (GCRs) represent the portion of CRs in the solar system that originates outside our galaxy's solar system [1]. Different astrophysical sources (e.g., active galactic nuclei and supernova stars) produce such CRs, usually named primary CRs (PCRs) to distinguish them from CRs that originate from the interactions of PCRs with other objects during their travel into the galaxy or within solar system regions (e.g., the interstellar medium and the planet's atmosphere), which are referred to as secondary CRs (SCRs) for that reason. Direct measurement of CR is one of the main tasks of many astroparticle experiments on Earth's surface and in space. Their measurements could illuminate many unsolved fundamental physics problems (e.g., dark matter composition and the existence of primordial antimatter). GCRs are composed of light particles, bare nuclei of different species, and a wide energy range from a few keV up to TeV; they can cause atoms they pass through to ionize. GCRs are of interest since they are present continuously in the beyond low Earth orbit (BLEO) region and can pass practically unimpeded through a typical spacecraft or the skin of an astronaut. This SR component is a primary concern for BLEO space missions involving long-duration travel, where radiation exposure levels can be several orders of magnitude higher than those experienced in low Earth orbit (LEO) or on

Earth's surface. Protons typically dominate the GCR composition, though the relative abundances of other particles can vary substantially.

In the past two decades, many astroparticle experiments have been built and deployed in space to investigate a number of open questions in fundamental physics and cosmology, such as the existence and composition of dark matter and energy and of primordial antimatter. A particular class of experiments, involving cosmic ray detectors (CRDs), is designed to produce a complete inventory of charged particles (CPs) and nuclei in CRs since the knowledge of this information is crucial to solving the above open problems in physics. The fundamental questions of CR physics relate to their origin, mechanism of acceleration to high energies, and their composition (i.e., the abundance of each nuclei particle) [1]. The principal CRDs operating in space include AMS-02, CALET, and ISS-CREAM, which are installed on the International Space Station (ISS), and ACE and DAMPE, which are based on satellite space missions.

3.2.1 *Principal operating CRDs on board the ISS*

3.2.1.1 *Alpha Magnetic Spectrometer*

The Alpha Magnetic Spectrometer (AMS) is a high-energy particle physics experiment in space, designed to measure CRs. The primary purposes of the experiment are the indirect search for dark matter through its annihilation products, the search for relic antimatter, and the precise measurement of the spectra of all CR species. Moreover, its capability to measure the variation in time of SR components makes it an optimal instrument for accurately assessing the space environment, which is necessary for improving tools and algorithms for risk assessment during space exploration (see Fig. 3.1).

The AMS-02 spectrometer consists of a permanent magnet and several instruments (subdetectors), a silicon tracker, a time-of-flight (TOF) sensor, a ring imaging Cherenkov counter (RICH), an electromagnetic calorimeter (ECAL), an anticoincidence counter (ACC), and a transition radiation detector (TRD). AMS-02 provides excellent particle identification capabilities. It measures the charge of the traversing particle independently in the tracker, RICH, and TOF subdetectors. The TOF and RICH subdetectors also measure particle velocity. AMS-02 was launched and

Fig. 3.1.   The AMS-02 was installed on the ISS on 19 May 2011. Immediately after its installation, it was powered up and began recording and transmitting data.

Source: Courtesy of AMS collaboration in "The Detector" (https://ams02.space/detector).

installed on the ISS in May 2011 and has been continuously operating since then [2]. The experiment is planned to be operational until 2030. Recently, the AMS collaboration has proposed a detector upgrade, that would increase the acceptance of the AMS detector by 300%. The upgrade is planned to be installed in 2026 [3].

Tables 3.1(a)–(c) report some information extracted from papers on the AMS collaboration [4–22].

The tables present a list of measurements conducted by the AMS collaboration, showcasing various particles such as electrons, protons, helium, lithium, beryllium, boron, carbon, nitrogen, oxygen, fluorine, neon, sodium, magnesium, silicon, sulfur, and iron. The measurements include absolute flux and time variation for different rigidity ranges over specific periods. The data offer valuable insights into the particle behavior and flux characteristics within the observed energy ranges. The numbers in square brackets in the "Rigidity Range" column correspond to the number of statistical bins present in the manuscript for the fluxes. Additionally, in the "Measurement Type" column, the numbers in round brackets indicate the duration of the measurements in either days (d) or Bartel rotations (b).

Table 3.1. Published AMS collaboration data until May 2023 related to absolute flux measurement analyses of (a) elementary particles and light nuclei ($Z <= 4$) and (b) medium and heavy nuclei ($Z > 4$) and (c) those related to "time variation" measurement analyses. All measurement periods started in May 2011.

(a)

Particle	Ref.	Rigidity Range (GV), [Bins]	Measurement Type	Period	Number of Events
Electron	[4]	1–350 [19]	Absolute flux	2012	6.8×10^6
Electron	[5]	0.5–1000 [74]	Absolute flux	2013	10.1×10^6
Electron	[6]	0.5–1400 [75]	Absolute flux	2017	28.1×10^6
Proton	[9]	1–1800 [72]	Absolute flux	2013	3.8×10^8
Proton	[16]	1–450 [57]	Absolute flux	2015	2.8×10^9
Helium (He)	[13]	1.92–3000 [68]	Absolute flux	2013	50×10^6
Helium (He)	[10]	1.92–3000 [68]	Absolute flux	2016	90×10^6
Lithium (Li)	[15]	1.92–3300 [67]	Absolute flux	2016	1.9×10^6
Beryllium (Be)	[15]	1.92–3300 [67]	Absolute flux	2016	0.9×10^6
Boron (B)	[15]	1.92–2600 [67]	Absolute flux	2016	2.6×10^6
Boron (B)	[22]	2.15–3300 [66]	Absolute flux	2021	1.8×10^{11}
Carbon (C)	[16]	1.92–3000 [68]	Absolute flux	2016	8.4×10^6
Carbon (C)	[22]	2.15–3000 [48]	Absolute flux	2021	1.8×10^{11}

(b)

Particle	Ref.	Rigidity Range (GV), [Bins]	Measurement Type	Period	Number of Events
Nitrogen (N)	[17]	2.15–3300 [66]	Absolute flux	2016	2.2×10^6
Oxygen (O)	[16]	2.15–3000 [67]	Absolute flux	2016	7.0×10^6
Oxygen (O)	[22]	2.15–3000 [48]	Absolute flux	2021	1.8×10^{11}
Oxygen (O)	[22]	2.15–3300 [66]	Absolute flux	2021	1.8×10^{11}
Fluorine (F)	[20]	2.15–2900 [48]	Absolute flux	2019	0.29×10^6
Fluorine (F)	[22]	2.15–3000 [48]	Absolute flux	2021	1.8×10^{11}
Neon (Ne)	[18]	2.15–3000 [66]	Absolute flux	2018	1.8×10^6
Neon (Ne)	[22]	2.15–3000 [48]	Absolute flux	2021	$1.8 \times 10^{11}*$
Sodium (Na)	[19]	2.15–3000 [48]	Absolute flux	2019	0.46×10^6
Magnesium (Mg)	[18]	2.15–3000 [66]	Absolute flux	2018	2.2×10^6
Magnesium (Mg)	[22]	2.15–3000 [48]	Absolute flux	2021	$1.8 \times 10^{11}*$
Silicon (Si)	[18]	2.15–3000 [66]	Absolute flux	2018	1.6×10^6

(Continued)

Table 3.1.   *(Continued)*

(b)

Particle	Ref.	Rigidity Range (GV), [Bins]	Measurement Type	Period	Number of Events
Silicon (Si)	[22]	2.15–3000 [48]	Absolute flux	2021	1.8×10^{11}*
Sulfur (S)	[22]	2.15–3000 [48]	Absolute flux	2021	0.38×10^6
Iron (Fe)	[21]	2.65–3000 [46]	Absolute flux	2019	0.62×10^6

(c)

Particle	Ref.	Rigidity Range (GV), [Bins]	Measurement Type	Period	Number of Events
Electron$^{(e+ + e-)}$	[7]	1–41.9 [10]	Time variation (4015-d)	2021	2.0×10^8
Electron$^{(e+ + e-)}$	[8]	0.5–49.33 [52]	Time variation (79-b)	2017	23.5×10^6
Proton$^{(p+ + p-)}$	[11]	1–100 [30]	Time variation (114-b)	2019	5.5×10^9
Helium (He)	[12]	1.92–60 [40]	Time variation (79-b)	2017	112×10^6
Helium (He)	[14]	1.71–100 [26]	Time variation (2824-d)	2019	7.6×10^8

Note: *The total cosmic-ray events collected by AMS in the first ten years of operation.

It is crucial to note that the actual capabilities of the astroparticle experiments are constantly evolving, not only due to hardware upgrades but also because of improved data analysis techniques. In other words, these astroparticle experiments rely on a scientific community that is continuously evolving in its scientific targets and remains at the forefront of technological developments. One exciting example of this evolution is represented by a recent study on the trapped particle components of SR using the AMS data [23].

3.2.1.2 *Calorimetric Electron Telescope*

The Calorimetric Electron Telescope (CALET) has been in operation since 2015 on the external platform of the ISS' Japanese experimental module (KIBO/JEM). The instrument is optimized to precisely study the properties of extreme-energy cosmic electrons up to several tens of TeV. It can also measure the relative composition and abundance of nuclei from space, ranging from protons to the heaviest elements up to $Z = 40$. CALET collected and sent over 1.8 billion CR events in the first three years of operation. The CALET calorimeter (CCAL) measures the total CR electron spectrum from ~1 GeV energy up to the TeV level. It comprises three instruments: a charge detector (CHD), a plastic scintillator hodoscope for absolute charge measurement, which is capable of detecting charges between 1 and ~40 Z, and two calorimeters (IMC+TACS) for CP energy measurements. It also includes a gamma-ray burst monitor (CGBM) sensitive to the soft X-ray (~7 keV) and the gamma-ray (~20 MeV) energy ranges. In addition to the primary instruments, there are two other main components: The advanced stellar compass (ASC) accurately determines the attitude in arcseconds, and the mission data controller (MDC) captures and formats the data from the instruments and sends the telemetry to the NASA ground station [24].

3.2.1.3 *Cosmic Ray Energetics and Mass for the ISS*

The Cosmic Ray Energetics and Mass for the International Space Station (ISS-CREAM) was successfully installed and activated on the Japanese Experiment Module Exposed Facility as an attached payload in 2017. This instrument is designed to study high-energy CRs by measuring the elemental spectra of nuclei ranging from hydrogen ($Z = 1$) up to iron ($Z = 26$) within an energy range from 1 to 1000 TeV. Additionally, it detects electrons at multi-TeV energies, complementing the capabilities of the original CREAM balloon-borne instrument [25].

Table 3.2 summarizes the characteristics of the CRDs operating on the ISS.

3.2.2 **Principal operating CRDs on board satellite space missions**

3.2.2.1 *Dark Matter Particle Explorer*

The Dark Matter Particle Explorer (DAMPE) is a satellite-based space mission whose primary purpose is the detection of cosmic electrons and photons up to energies of 10 TeV. The DAMPE instruments can also measure the fluxes and the elemental composition of the galactic CR nuclei up to 100 TeV. It has been collecting data since 2015. It consists of a double-layer plastic scintillator detector, a silicon-tungsten tracker converter, an ECAL, and a neutron detector. In the first years of operation, DAMPE has collected and sent more than 6 billion CR events [26].

Table 3.2. Cosmic ray detectors (CRDs) in space on board the ISS: a comparison of instruments and energy ranges.

CRD	Period of Activity	Main Instruments	Energy Range (Electron) GeV
AMS-02	2011–present	Transition radiation detector (TRD), electromagnetic calorimeter (ECAL), ring imaging Cherenkov counter (RICH), anticoincidence counter (ACC), time-of-flight (TOF) sensor	$0.5\text{–}3 \times 10^3$
CALET	2015–present	Calorimeter (CCAL), gamma-ray burst monitor (CGBM)	$1\text{–}10 \times 10^3$
ISS-CREAM	2017–present	TOF sensor, silicon charge detector (SCD), TRD	$10^3\text{–}10^6$

3.2.2.2 *Advanced Composition Explorer*

The Advanced Composition Explorer (ACE) is a satellite-based space mission that started its operation in 1998, intending to observe particles of solar, interplanetary, interstellar, and galactic origins, spanning the energy range from solar wind ions to galactic CR nuclei. It is located at the L1 Lagrange point, about 1.4 million kilometers from the Earth. ACE can measure particle and nuclei elements up to $Z = 30$ in an energy range of up to hundreds of MeV and is composed of nine different instruments.

The Cosmic-Ray Isotope Spectrometer (CRIS) on the ACE can cover an energy interval of 50–500 MeV/nucleon, with an isotopic resolution for elements of $Z = 2\text{–}30$. The nuclei detected in this energy interval are predominantly CRs originating in our galaxy. Charge and mass identification with CRIS is based on multiple measurements of dE/dx and total energy in stacks of silicon detectors and trajectory measurements in a scintillating optical fiber trajectory (SOFT) hodoscope. The instrument has a geometric factor of 250 cm²-sr for isotope measurements [27].

The Ultra-Low-Energy Isotope Spectrometer (ULEIS) on the ACE spacecraft is an ultrahigh-resolution mass spectrometer that measures particle composition and energy spectra of elements He–Ni with energies from ~45 keV/nucleon to a few MeV/nucleon [28].

Other ACE instruments also provide near-real-time solar wind information over short periods. When reporting space weather, ACE can provide a warning (about one hour) of geomagnetic storms that can overload power grids, disrupt communications on Earth, and pose a hazard to astronauts. An instrument used for such purposes is the Electron, Proton, and Alpha-Particle Monitor (EPAM), designed to measure a broad range of energetic particles over nearly the full unit sphere at high time resolutions. Such measurements of ions and electrons in the range of tens of keV to several MeV are essential for understanding the dynamics of solar flares (SFs), co-rotating interaction regions, interplanetary shock acceleration, and upstream terrestrial events. Another instrument is the Real-Time Solar Wind system, which uses low-energy energetic particles to warn of interplanetary shocks approaching and to help monitor the flux of high-energy particles that can cause radiation damage in satellite systems [29]. Thanks to careful orbital maintenance and fuel conservation, ACE is expected to operate through at least 2027, with extended fuel projections possibly allowing its mission to continue until 2029.

3.2.2.3 *Payload for Antimatter Matter Exploration and Light-Nuclei Astrophysics*

The Payload for Antimatter Matter Exploration and Light-Nuclei Astrophysics (PAMELA). PAMELA operations started in 2006, producing accurate measurements of the CR components (particle and light nuclei up to $Z = 6$) [30]. Launched on 15 June 2006 by a Russian Soyuz rocket, it was a groundbreaking satellite-based experiment. Developed through an international collaboration, mainly between Italian and Russian scientific institutions, its primary goal was to study CRs, particularly antimatter, in search of clues about dark matter. PAMELA carried a sophisticated suite of detectors designed to measure CRs with unprecedented precision. The key components included a silicon tracker, which could identify particles' charge and momentum, and a silicon-tungsten calorimeter, crucial for measuring the energy of incoming particles and distinguishing between different types of radiation. It also hosted a TOF system for determining particles' velocity and a neutron detector to capture secondary particles produced by cosmic-ray interactions. These detectors allowed PAMELA to measure the fluxes of positrons, electrons, protons, and antiprotons across a wide energy range, providing invaluable data. One of the mission's most significant findings was the first unexpected observation of an excess of positrons, confirmed by the AMS-02 detector, which challenged existing models and suggested potential dark matter interactions or unknown astrophysical sources. PAMELA's success set the stage for future missions similar to AMS-02, as previously described.

Table 3.3 summarizes the characteristics of the CRDs operating on the satellite space missions.

Table 3.3. Cosmic ray detectors in satellite space missions: a comparison of instruments and energy ranges, CRD operations, and measurements.

CRD	Location	Period of Activity	Main Instruments	Energy Range (Electron) GeV
DAMPE	Sun-sync orbit satellite 500 km	2015–present	Plastic scintillator detector (PSD), tracker, Calorimetric Electron Telescope (ECAL), neutron detector (ND)	$5–100 \times 10^3$
ACE	Lissajous orbit satellite 1.5×10^6 km	1998–present	Cosmic-Ray Isotope Spectrometer (CRIS), Electron, Proton, and Alpha-Particle Monitor (EPAM), Magnetometer (MAG), Real-Time Solar Wind (RTSW), Solar Energetic Particle Ionic Charge Analyzer (SEPICA), Solar Isotope Spectrometer (SIS), Solar Wind Electron, Proton, and Alpha Monitor (SWEPAM), Solar Wind Ion Composition Spectrometer (SWICS), Solar Wind Ion Mass Spectrometer (SWIMS), Ultra-Low-Energy Isotope Spectrometer (ULEIS)	0.05–0.5
PAMELA	Polar orbit 300–600 km	2006–2021	Tracker, ECAL, Time-of-flight (TOF) sensor	$50–2 \times 10^3$

3.2.3 *CRD operations and measurements*

CRDs operations in space, particularly those with missions lasting several years, offer a unique opportunity to enhance our understanding of the potential health effects of IR on humans in space. By collecting vast amounts of data over extended periods, these detectors provide insights into cosmic radiation that are critical for the safety and well-being of astronauts. A CRD annually registers over a billion CR events. For instance, since the start of data collection in 2011, AMS-02 has recorded more than 190 billion events. Another notable example is the ACE, which has been operational since its launch in 1997 and has provided valuable data on CRs and solar wind particles. The following characteristics of CRD operations are particularly relevant to advancing our knowledge in this area [31]:

- *Complete CR component identification*: CRDs excel in measuring the abundances and spectral distribution of various CR particles, from protons and electrons to heavier nuclei such as helium and iron ($Z = 26$), with unprecedented precision and accuracy.
- *High-energy range spectra*: The data collected by CRDs spans an impressive energy range, from a few MeV to hundreds of TeV.
- *CR solar modulation*: CRDs are crucial for studying variations in CR flux over time, particularly in relation to solar activity and cycles. These variations, especially during Solar Cycles 23 and 24 and potentially into the 25th, are essential for understanding the effects of IR on space missions.

3.2.3.1 *Particle identification capability*

One of the most significant features of modern CRDs is their advanced particle identification capabilities. CRDs are designed to differentiate between particles that constitute CRs, including protons, electrons, and a wide range of atomic nuclei. This identification process is critical because each type of particle interacts with matter, including human tissue, in distinct ways, leading to different health risks for astronauts.

The ability to identify particles with such high precision stems from sophisticated detection technologies, such as silicon tracker detectors, TOF systems, and calorimeters. These technologies work in concert to measure each incoming particle's charge, mass, and energy. For instance,

AMS-02's magnetic spectrometer uses a strong magnetic field to bend the paths of CPs, allowing scientists to determine their momentum and charge. Similarly, the ACE mission employs instruments such as the CRIS and Solar Isotope Spectrometer (SIS), which measure the charge, mass, and energy of incoming particles at high resolutions. This level of detail enables CRDs to distinguish between lighter particles, such as protons and helium nuclei, and heavier elements, such as iron. The precision in particle identification enhances our understanding of CR composition and allows for the accurate assessment of the radiation dose to which astronauts are exposed during space missions.

Moreover, this capability is essential for studying rare particles such as antiprotons and positrons, which could provide clues about dark matter and other fundamental questions in astrophysics. The ability to precisely identify and measure these particles' energy spectra allows researchers to explore CR sources and propagation mechanisms, thereby contributing to our broader understanding of the universe.

3.2.3.2 *Energy range and spectral analysis*

Another critical feature is the energy range over which CRDs can detect CRs. The detectors can measure particles across a broad spectrum of energies, from a few MeV to several TeV. This extensive energy range is essential for several reasons.

First, different energy levels of CRs interact with spacecraft materials and human tissues differently. Spacecraft hulls are more easily shielded from low-energy particles but can still present significant risks during extended space missions. Conversely, less frequent high-energy particles are more penetrating and can potentially cause more severe biological damage. Understanding the full spectrum of CR energies is crucial for designing effective shielding and assessing the long-term health risks to astronauts.

CRDs such as AMS-02 and ACE are equipped with instruments that allow them to measure the energy of incoming particles with high accuracy. AMS-02 uses a combination of a magnetic spectrometer and a silicon tracker to determine the momentum and charge of CR particles across a wide energy range. On the other hand, ACE employs instruments such as CRIS, which is particularly adept at measuring the isotopic composition of CRs in the energy range from 50 MeV/nucleon to several GeV/nucleon. This broad energy spectrum allows ACE to study the origins and propagations of CRs, as well as their interactions with the solar wind.

The comprehensive energy spectrum provided by CRDs is also invaluable for understanding variations in CR intensity over time, particularly in response to solar activity. By analyzing the spectral distribution of CRs, scientists can gain insights into the processes governing their propagation through the galaxy, including interactions with interstellar matter and magnetic fields. Furthermore, these data are crucial for predicting increased radiation risk during space missions, contributing to more effective mission planning and crew safety protocols.

3.2.3.3 *Continuous monitoring across solar cycles*

One of the most significant advantages of long-duration CRD missions is their ability to monitor CRs continuously across different solar cycles. Solar cycles, which typically last about 11 years, involve variations in the Sun's magnetic activity that affect the intensity and composition of CRs reaching Earth and space environments.

The solar modulation of CRs is a complex phenomenon where the solar wind, a stream of CPs emitted by the Sun, interacts with GCRs, leading to variations in their intensity. During periods of high solar activity (solar maximum), the increased solar wind and magnetic fields can deflect many lower-energy CRs, reducing their intensity. Conversely, during solar minimum, the weaker solar wind allows more CRs to penetrate the solar system, increasing their flux.

CRDs such as AMS-02 and ACE have been operational during Solar Cycles 23 and 24 and are expected to continue monitoring into Cycle 25. This continuous monitoring is crucial for understanding how CR flux changes and how these changes might impact astronauts on long-duration missions. For example, since its launch, ACE has provided near-continuous data on CRs, solar wind, and interstellar particles, offering invaluable insights into how solar activity influences CR flux and radiation exposure in space. The data collected can help predict periods of higher radiation exposure, allowing for better planning of spacewalks and other mission activities that involve increased risk.

Furthermore, the ability to monitor CRs over multiple solar cycles provides insights into long-term trends in solar modulation and its effects on CRs. This information is critical for space missions and understanding the broader impacts of CRs on Earth's atmosphere and climate. For example, variations in CR intensity can influence cloud formation and weather patterns, making CRDs essential tools for studying space weather and its potential effects on Earth's environment.

The variation of CR flux over time is crucial for investigating the effects of IR during space missions. Figure 3.2 shows proton flux variation at low energies as reconstructed by AMS-02 since May 2011.

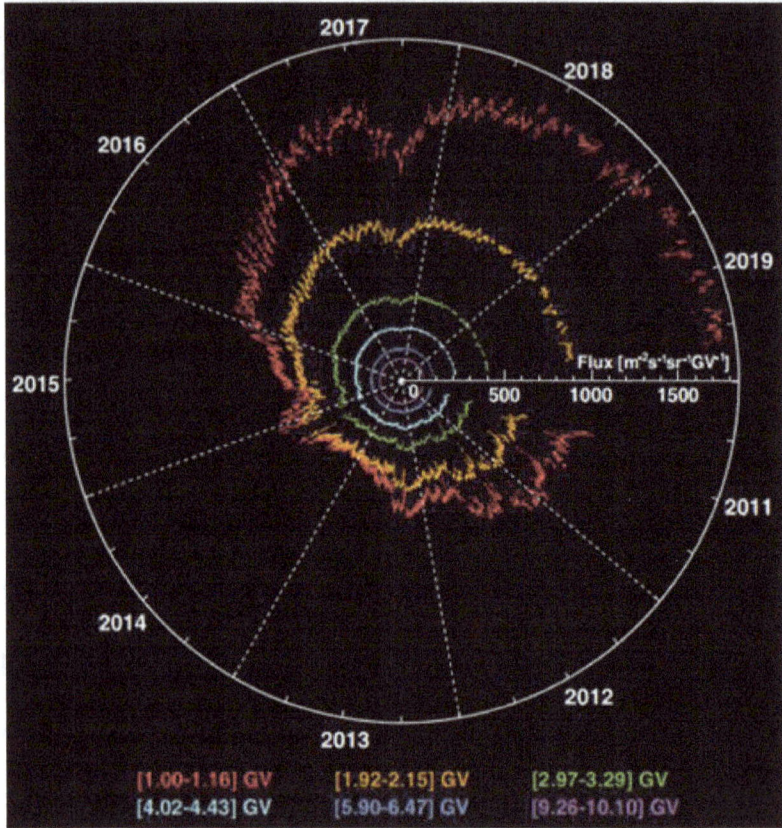

Fig. 3.2. The daily AMS proton fluxes for six typical rigidity bins from 1.00 to 10.10 GV, measured from 20 May 2011 to 29 October 2019, including a significant portion of solar cycle 24 (from December 2008 to December 2019). The AMS data cover the ascending, maximum, and descending phases up to the minimum of Solar Cycle 24. Days with SEPs are removed for the two lowest rigidity bins. The gaps in the fluxes are due to detector studies and upgrades. The error bars are invisible. As seen, the proton fluxes exhibit significant variations with time, and the relative magnitude of these variations decreases with increasing rigidity.

Source: Courtesy of AMS Collaboration in Bartoloni, A. *et al.* (2024). Bridging the Gap: Exploring AMS, Astroparticle Experiments and Space Radiobiology. *Proceedings of Science*, Volume 44921, European Physical Society Conference on High Energy Physics. CC BY 4.0 International (https://creativecommons.org/licenses/by/4.0/).

3.2.3.4 *Characterization of the Van Allen belts*

The Van Allen radiation belts are regions of CPs trapped by Earth's magnetic field, primarily composed of energetic electrons and protons, with a smaller fraction of heavier ions. These belts are structured into two distinct zones: the inner belt, extending from approximately 1,000 to 12,000 km, and the outer belt, ranging from 13,000 to 60,000 km above Earth's surface.

A significant localized feature within the belts is the South Atlantic Anomaly (SAA), where the inner belt dips unusually close to Earth's surface, down to altitudes as low as 200 km. This results in elevated radiation levels that pose increased risks to spacecraft systems and astronauts, particularly in LEO, where radiation-induced electronic malfunctions and biological hazards become more prominent.

Initially discovered in 1958 by Van Allen and his team, the belts were identified as zones containing energetic electrons and ions confined by the geomagnetic field. Early empirical models such as NASA's AE-8 and AP-8 were developed in the 1960s–1970s and remain in use for spacecraft design [31]. These models, however, assume quasi-static conditions that do not reflect the belts' now well-established variability.

Recent studies have emphasized the dynamic nature of the radiation belts, particularly the outer zone [32]. During geomagnetic storms, fluxes of high-energy electrons can vary by several orders of magnitude within timescales ranging from minutes to days. Notably, an analysis of 276 geomagnetic storms across a solar cycle revealed that the belts' response is non-uniform, with some storms enhancing, suppressing, or leaving the radiation environment unchanged [33]. This complexity arises from a delicate balance between particle acceleration and loss mechanisms.

Advanced CRDs on board space missions, such as AMS and PAMELA, offer three-dimensional (3D) mapping capabilities that allow precise characterization of the belts' particle populations. These detectors measure flux, energy, and species composition across various orbital locations, enabling spatially resolved radiation models. This capability is particularly valuable for profiling high-risk zones like the SAA and the outer belt during disturbed geomagnetic conditions.

Data from CRDs have revealed the presence of not only electrons and protons but also more energetic and massive ions, which represent serious threats to electronics and crewed missions. The 3D reconstruction of the belts facilitates accurate risk assessment and the development of optimized shielding strategies for future missions.

Such data are increasingly vital for BLEO missions, including pro-posed cis-lunar space stations such as the Lunar Gateway. Understanding the belts' structure and variability is essential for trajectory planning, habitat shielding design, and crew safety, especially in environments where Earth's magnetic protection is limited or absent.

In addition, real-time radiation monitoring by CRDs supports space weather forecasting, allowing mission operators to implement counter-measures during solar storms, when radiation levels in the belts can rap-idly escalate. Overall, the integration of high-resolution CRD data into spacecraft design and mission planning represents a critical advancement in ensuring the safety and success of human and robotic exploration beyond Earth orbit.

3.2.3.5 *Potential applications of CRD data for space radiobiology*

The importance of CRDs in space extends beyond their primary mission of measuring cosmic radiation. The data collected by these detectors can also be harnessed for space radiobiology, a field that investigates the effects of SR on biological organisms, particularly human health. Understanding the impact of GCRs and SPEs on human tissue is critical for planning long-duration space missions, especially those venturing BLEO into deep space.

A case study of a preliminary evaluation of this potential application can be drawn from the capabilities demonstrated by the AMS-02 detector [34]. This research aimed to identify the components of cosmic radiation most relevant for assessing risks associated with crewed exploratory space missions, both in LEO and beyond. Existing studies on SR sensitivity analysis [35] served as a foundation for this work [36, 37], and it became evident that the energy range critical for radiation risk assessment closely aligns with the measurement capabilities of the AMS-02 detector.

AMS-02 and other CRDs have proven capable of precisely measuring CPs and heavy nuclei (HNs) within the energy ranges pertinent to human health risk assessments. For instance, the energy spectrum that is particu-larly hazardous to human health, such as high-energy protons and heavy ions, is well within the detection range of AMS-02. This alignment between the CRDs' measurement capabilities and the energy ranges of interest for space radiobiology underscores the potential of these instruments to contribute valuable data for evaluating radiation risks to astronauts.

A significant aspect of space radiobiology is understanding how different types of SR, primarily GCRs and SPEs, interact with human tissue. These interactions can lead to various health risks, including increased cancer risk, central nervous system effects, and acute radiation syndrome during high-dose exposure events. The data provided by CRDs, including AMS-02, can be instrumental in modeling these interactions, allowing researchers to predict the biological effects of SR more accurately [38, 39].

Also, these potentialities were successively evaluated in a research study [40], further confirming the significance of CRDs in measuring CPs and HNs within the energy range crucial for assessing the human radiation hazard. These findings demonstrate that these instruments provide valuable data for basic CR research and are directly applicable to assessing the radiation environment's impact on human health.

Moreover, the continuous data stream provided by CRDs offers a temporal dimension to space radiobiology studies. This aspect is crucial because the SR environment is highly variable and influenced by solar cycles, geomagnetic activity, and other factors. By correlating the radiation data from CRDs with biological studies, researchers can better understand how fluctuations in radiation exposure levels might affect astronauts over time. For example, during solar particle events, CRDs can capture the surge in radiation levels, which can then be used to evaluate the potential acute and chronic health risks posed by such events.

In a broader sense, the data from CRDs can support the development of more accurate risk models for space missions, informing the design of spacecraft shielding, mission planning, and crew health monitoring strategies. These models could be refined to consider individual variability in radiation sensitivity, which could become increasingly important as missions grow longer and more ambitious, such as those planned for Mars exploration.

In conclusion, while CRDs such as AMS-02 were primarily designed for CR research, their data have proven to be valuable resources for space radiobiology. By providing detailed information on the energy spectra and fluxes of particles that pose health risks to astronauts, CRDs play a crucial role in the ongoing efforts to safeguard human health during space missions. This preliminary evaluation suggests that further integration of CRD data into radiobiological research could significantly enhance our understanding of SR risks and improve the safety of future crewed missions in deep space (see Fig. 3.3).

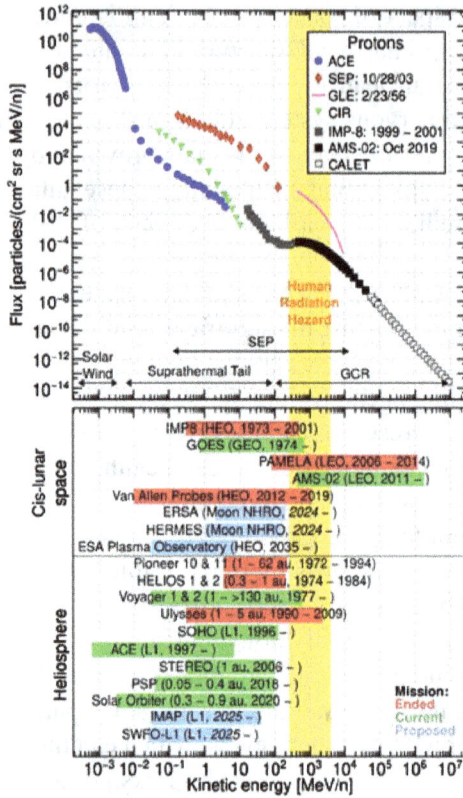

Fig. 3.3. Flux vs. energy ranges of protons as seen by different CRDs and probes operating over the past decades or those planned soon. The yellow band range is the most relevant for human radiation exposure.

Source: Courtesy of F. Corti *et al.* (2022). Galactic CRs and solar energetic particles in cis-lunar space: Need for contextual energetic particle measurements at Earth and supporting distributed observation. CC BY 4.0 International (https://creativecommons.org/licenses/by/4.0/).

3.2.4 *Next-generation CRDs*

The next generation of CRDs represents a significant leap forward in our ability to study high-energy particles and cosmic phenomena. These advanced instruments are designed to measure CRs across a broad spectrum of energies, providing critical insights into the origins of CRs, the nature of dark matter, and the potential existence of primordial antimatter.

One critical point of these projects is the reach of higher acceptance with respect to previous-generation CRDs.

3.2.4.1 *Acceptance in cosmic-ray detectors*

Acceptance is one of the key metrics in the design and effectiveness of CRDs. In CRDs, acceptance is defined as the product of the detector's geometric area and field of view, typically expressed in square meters steradian (m² sr). It represents the detector's ability to collect incoming particles over a given area and solid angle, thus determining how much data it can gather about CRs over time. Higher acceptance means the detector can capture more particles, leading to improved statistical accuracy and the ability to study rarer events, such as detecting antimatter particles such as anti-helium. In space missions, improving acceptance is crucial because it directly impacts the amount and quality of data collected, especially in missions with long durations and distant locations, such as ALADInO at the Earth–Sun L2 Lagrangian point. Table 3.4 reports a comparison of the acceptance of various CRDs.

The following paragraphs describe three futuristic key projects of large-acceptance CRDs: HERD, designed to operate on the CSS, ALADInO, and AMS-100, where a completely different approach is used to design the next-generation magnetic spectrometer in space. In such cases, an improvement is made in the planned geometrical acceptance, but the operational location is entirely new and disruptive since it is moved to the Sun–Earth Lagrange Point 2, 1.5 million km from Earth, opposite the Sun's direction.

3.2.4.2 *HERD*

The High-Energy Cosmic Radiation Detection Facility is a cutting-edge project to advance our understanding of cosmic and gamma rays in LEO. HERD is designed to detect CRs within an energy range from a few GeV to hundreds of TeV. The core of HERD's instrumentation is its 3D calorimeter, which consists of a stack of thin scintillating fibers interleaved with tungsten plates. This design allows HERD to measure the energy and direction of incoming particles with high precision. The large acceptance area of HERD, approximately 3 m² sr, enables it to collect a vast amount of data, thereby improving the statistical accuracy of CR measurements.

Table 3.4. Past, present, and future CRD acceptance values.

CRD	Acceptance (m² sr)	Comments
	Past and Present CRDs	
AMS-02	Full: 0.05 Inner: 0.5	Installed on the ISS since 2011; uses a permanent magnet.
CALET	0.12	Calorimetric Electron Telescope on the ISS; focused on high-energy electron and gamma-ray detection.
ISS-CREAM	0.6	Cosmic Ray Energetics and Mass experiment; balloon-borne version had higher acceptance.
PAMELA	0.002	Satellite-based; focused on antimatter detection, particularly positrons and antiprotons.
ACE	0.0015	Advanced Composition Explorer; placed at L1 Lagrange Point; specialized in isotopic measurements.
DAMPE	0.3	Dark Matter Particle Explorer; satellite mission focused on high-energy electron and gamma-ray detection.
	Future CRDs	
HERD	~3	Future CRD to be placed on the Chinese Space Station; 3D calorimeter with large acceptance.
ALADInO	(a) MS only: >10 (b) Calo only: 9 (c) Combined: 3	Future CRD at Earth–Sun L2 Lagrange Point; optimized for antimatter detection.
AMS-100	~20	Proposed for BLEO; would significantly improve AMS-02's capabilities.

Additionally, HERD is equipped with silicon tracker planes to determine particle trajectories and charge identification modules to distinguish between different particle types. HERD is in the advanced development stage, with international collaboration primarily led by China. It is expected to be launched and deployed on the Chinese Space Station within the next few years [41].

3.2.4.3 *ALADInO*

The Antimatter Large Acceptance Detector In Orbit (ALADInO) is a novel CRD designed to measure high-energy CRs and search for antimatter components in space. This project is poised to succeed the PAMELA and AMS-02 missions by focusing on detecting rare antimatter particles, such as antiprotons, antideuterons, and antihelium. ALADInO will operate at the Sun–Earth L2 Lagrangian point, providing a stable environment far from Earth's magnetic field [42].

The ALADInO detector features a superconducting magnet, a high-precision tracking system, and a TOF system. This configuration allows ALADInO to achieve a maximum detectable rigidity (MDR) of over 20 TVs, which is crucial for separating matter from antimatter. The inner 3D imaging calorimeter enhances its ability to accurately measure CRs up to PeV energies. ALADInO is still in the conceptual and design phase, with ongoing research focused on refining its instrumentation and ensuring its feasibility for long-term space missions [42].

In the following, we compare ALADInO with AMS-02:

Location and operation: AMS-02 has been operating on the ISS since 2011 and is expected to extend its mission beyond 2030. In contrast, ALADInO will be deployed at the Earth–Sun L2 Lagrangian point, which offers a more stable and isolated environment, ideal for high-precision measurements. ALADInO is projected to begin operations in the mid-2040s and will have a mission duration exceeding five years.

Acceptance: The acceptance area, which indicates the detector's capability to capture CRs, is another area where ALADInO shows significant advancements. The ALADInO detector's acceptance is expected to exceed 10 m^2 sr when using its magnetic spectrometer alone and 9 m^2 sr with the calorimeter alone. When combining both systems, the acceptance remains around 3 m^2 sr. This is a vast improvement over AMS-02, which has a complete acceptance of 0.05 m^2 sr and an inner acceptance of 0.5 m^2 sr.

Magnetic field and MDR: ALADInO is designed with a superconducting magnet, providing an average magnetic field of 0.8 Tesla (T), significantly higher than AMS-02's 0.15 T permanent magnet. This enhancement allows ALADInO to achieve an MDR of over 20 TV, compared to AMS-02's 2 TV for $Z = 1$ particles and 3.2 TV for particles with $Z > 1$.

The higher MDR of ALADInO enables the detection of particles at much higher energies, which is crucial for identifying rare antimatter components such as antihelium.

Calorimeter depth and energy resolution: Another critical improvement involves ALADInO's calorimeter, which has a total depth of 61 radiation lengths (X_0) and 3.5 interaction lengths (λI), compared to AMS-02's 17 X_0 and 0.6 λI. This deeper calorimeter enhances ALADInO's ability to measure high-energy particles accurately. Additionally, ALADInO's energy resolution for electrons and positrons is expected to be 1.5%, compared to AMS-02's 2%, providing more precise measurements.

Particle separation and channels: ALADInO also promises superior particle separation capabilities, crucial for distinguishing electrons and protons. Its e/p separation efficiency is expected to exceed that of AMS-02, which has an e/p separation efficiency of 25% at 1 m^2 sr and 35% at 5 m^2 sr. Furthermore, ALADInO will have over 10^6 readout channels, compared to AMS-02's 10^4–10^5 channels, allowing for finer spatial resolution and better data quality.

Mass and power: Despite these advancements, ALADInO is designed to be more power-efficient, operating on less than 2.5 kW of power compared to AMS-02's 3.0 kW. ALADInO's total mass is expected to be around 7.5 tons, slightly heavier than AMS-02's 6.5 tons, reflecting the additional instrumentation required for its enhanced capabilities.

Overall, ALADInO will represent a significant technological leap from AMS-02, providing a more powerful tool for probing the CRs' highest energies and searching for elusive antimatter. This will make it a pivotal project for the future of astroparticle physics [42].

3.2.4.4 *AMS-100*

The Alpha Magnetic Spectrometer 100 (AMS-100) represents a significant evolution of the AMS-02 experiment operating on the ISS. This proposed detector aims to be deployed on the BLEO, potentially on a deep-space platform. AMS-100 is designed to increase sensitivity by 100 times compared to AMS-02, enabling it to detect ultrarare CR events and probe deeper into the cosmic radiation spectrum. AMS-100's instrumentation includes an advanced magnetic spectrometer with improved tracking capabilities and a larger acceptance area, allowing for the detection of CRs and antimatter with unprecedented precision. It also incorporates state-of-the-art calorimeters and Cherenkov detectors to measure particle energy and identify particle types. The magnet design is based on high-temperature superconducting tapes, which allow the construction of a thin solenoid with a homogeneous magnetic field of 1 T inside. The inner volume will be instrumented with a silicon tracker with a maximum detectable rigidity of 100 TV and a calorimeter system that is 70 radiation lengths deep, equivalent to four nuclear interaction lengths, which extends the energy reach for CR nuclei up to the PeV scale, i.e., beyond the CR knee.

Currently, AMS-100 is in the proposal and preliminary design phase, with extensive international collaboration focusing on securing the necessary funding and technological development to bring this ambitious project to fruition [43].

References

[1] Gaisser, T. K., Engler, R., and Resconi, E. (2016). *Cosmic Rays and Particle Physics — Second Edition.* Cambridge University Press, Cambridge, UK.

[2] Aguilar, M., Cavasonza, L. Ali., Ambrosi, L., Arruda, G., Attig, L. N., Barao, F., Barrin, L., Bartoloni, A., Başeğmez-du Pree, S., *et al.* (2021). The Alpha Magnetic Spectrometer (AMS) on the International Space Station: Part II — Results from the first seven years. *Physics Reports*, 894, 1–116. https://doi.org/10.1016/j.physrep.2020.09.003.

[3] Bollweg, K. on behalf of the AMS Collaboration (2022). Upgrade of AMS for the next Ten Years. *Conference Proceedings of the COSPAR 2022 44th Scientific Assembly*, 16–24 July 2022, Athens, Greece.

[4] Aguilar, M., *et al.* (AMS Collaboration). (2013). First result from the Alpha Magnetic Spectrometer on the International Space Station: Precision measurement of the positron fraction in primary cosmic rays of 0.5–350 GeV. *Physical Review Letters*, 110, 141102. https://doi.org/10.1103/PhysRevLett.110.141102.

[5] Aguilar, M., *et al.* (AMS Collaboration). (2014). Precision measurement of the (e+ + e−) flux in primary cosmic rays from 0.5 GeV to 1 TeV with the Alpha Magnetic Spectrometer on the International Space Station. *Physical Review Letters*, 113, 221102. https://doi.org/10.1103/PhysRevLett.113.221102.

[6] Aguilar, M., *et al.* (AMS Collaboration). (2019). Towards understanding the origin of cosmic-ray electrons. *Physical Review Letters*, 122, 101101. https://doi.org/10.1103/PhysRevLett.122.101101.

[7] Aguilar, M., *et al.* (AMS Collaboration). (2023). Temporal structures in electron spectra and charge sign effects in galactic cosmic rays. *Physical Review Letters*, 130, 161001. https://doi.org/10.1103/PhysRevLett. 130.161001.

[8] Aguilar, M., *et al.* (AMS Collaboration). (2018). Observation of complex time structures in the cosmic-ray electron and positron fluxes with the Alpha Magnetic Spectrometer on the International Space Station. *Physical Review Letters*, 121, 051102. https://doi.org/10.1103/PhysRevLett.121. 051102.

[9] Aguilar, M., *et al.* (AMS Collaboration). (2015). Precision measurement of the proton flux in primary cosmic rays from rigidity 1 GV to 1.8 TV with the Alpha Magnetic Spectrometer on the International Space Station. *Physical Review Letters*, 114, 171103. https://doi.org/10.1103/PhysRevLett. 114.171103.

[10] Aguilar, M., *et al.* (AMS Collaboration). (2016). Antiproton flux, antiproton-to-proton flux ratio, and properties of elementary particle fluxes in primary cosmic rays measured with the Alpha Magnetic Spectrometer on the International Space Station. *Physical Review Letters*, 117, 091103. https://doi.org/10.1103/PhysRevLett.117.091103.

[11] Aguilar, M., *et al.* (AMS Collaboration). (2021). Periodicities in the daily proton fluxes from 2011 to 2019 measured by the Alpha Magnetic Spectrometer on the International Space Station from 1 to 100 GV. *Physical Review Letters*, 127, 271102. https://doi.org/10.1103/PhysRevLett.127. 271102.

[12] Aguilar, M., *et al.* (AMS Collaboration). (2018). Observation of fine time structures in the cosmic proton and helium fluxes with the Alpha Magnetic Spectrometer on the International Space Station. *Physical Review Letters*, 121, 051101. https://doi.org/10.1103/PhysRevLett.121.051101.

[13] Aguilar, M., *et al.* (AMS Collaboration). (2015). Precision measurement of the helium flux in primary cosmic rays of rigidities 1.9 GV to 3 TV with the Alpha Magnetic Spectrometer on the International Space Station. *Physical Review Letters*, 115, 211101. https://doi.org/10.1103/PhysRevLett. 115.211101.

[14] Aguilar, M., *et al.* (AMS Collaboration). (2022). Properties of daily helium fluxes. *Physical Review Letters*, 128, 231102. https://doi.org/10.1103/ PhysRevLett.128.231102.

[15] Aguilar, M., *et al.* (AMS Collaboration). (2018). Observation of new properties of secondary cosmic rays (SCRs) Lithium, Beryllium, and Boron by the Alpha Magnetic Spectrometer on the International Space Station.

Physical Review Letters, 120, 021101. https://doi.org/10.1103/PhysRevLett. 120.021101.

[16] Aguilar, M., *et al.* (AMS Collaboration). (2017). Observation of the identical rigidity dependence of He, C, and O cosmic rays at high rigidities by the Alpha Magnetic Spectrometer on the International Space Station. *Physical Review Letters*, 119, 251101. https://doi.org/10.1103/ PhysRevLett.119.251101.

[17] Aguilar, M., *et al.* (AMS Collaboration). (2018). Precision measurement of cosmic-ray nitrogen and its primary and secondary components with the Alpha Magnetic Spectrometer on the International Space Station. *Physical Review Letters*, 121, 051103. https://doi.org/10.1103/PhysRevLett. 121.051103.

[18] Aguilar, M., *et al.* (AMS Collaboration). (2020). Properties of neon, magnesium, and silicon primary cosmic rays results from the Alpha Magnetic Spectrometer. *Physical Review Letters*, 124, 211102. https://doi.org/10.1103/ PhysRevLett.124.211102.

[19] Aguilar, M., *et al.* (AMS Collaboration). (2021). Properties of a new group of cosmic nuclei: Results from the Alpha Magnetic Spectrometer on sodium, aluminum, and nitrogen. *Physical Review Letters*, 127, 021101. https://doi.org/10.1103/PhysRevLett.127.021101.

[20] Aguilar, M., *et al.* (AMS Collaboration). (2021). Properties of heavy secondary fluorine cosmic rays: Results from the Alpha Magnetic Spectrometer. *Physical Review Letters*, 126, 081102. https://doi.org/10.1103/ PhysRevLett.126.081102.

[21] Aguilar, M., *et al.* (AMS Collaboration). (2021). Properties of iron primary cosmic rays: Results from the Alpha Magnetic Spectrometer. *Physical Review Letters*, 126, 041104. https://doi.org/10.1103/PhysRevLett.126. 041104.

[22] Aguilar, M., *et al.* (AMS Collaboration). (2023). Properties of cosmic-ray sulfur and determination of the composition of primary cosmic-ray carbon, neon, magnesium, and sulfur: Ten-year results from the alpha magnetic spectrometer. *Physical Review Letters*, 130, 211002. https://doi.org/10.1103/ PhysRevLett.130.211002.

[23] Aguilar, M., *et al.* (2022). Properties of heavy nuclei in South Atlantic anomaly region. *Conference Proceedings of the COSPAR 2022 44th Scientific Assembly*, 16–24 July 2022, Athens, Greece.

[24] Asaoka, Y., Adriani, O., Akaike, Y., Asano, K., Bagliesi, Berti, M. G., Bigongiari, E., G., Binns, W. R., Bonechi, S., Bongi, M., and *et al.* (2019). The CALorimetric Electron Telescope (CALET) on the International Space Station: Results from the first two years on orbit. *Journal of Physics: Conference Series*, 1181. https://doi.org/10.1088/1742-6596/1181/1/ 012003.

[25] Seo, E. S., Anderson, T., Angelaszek, D., Baek, S. J., Baylon, J., Buénerd, M., Copley, M., Coutu, S., Derome, L., Fields, B., and *et al.* (2014). Cosmic Ray Energetics and Mass for the International Space Station (ISS-CREAM). *Advances in Space Research*, 53, 1451–1455. https://doi.org/10.1016/j.asr.2014.01.013.

[26] De Benedittis, A. for the DAMPE Collaboration. (2019). The DAMPE experiment: Performance and Results. *Proceedings of EPS-HEP 2019.*

[27] Stone, E. C., Frandsen, A. M., Mewaldt, R. A., Christian, E. R., Margolies, D., Ormes, J. F., Snow, F. (1998). The advanced composition explorer. *Space Science Reviews*, 86, 1–22. https://doi.org/10.1023/A:1005082526237.

[28] Gold, R. E., Krimigis, S. M., Hawkins III, S. E., Haggerty, D. K., Lohr, D. A., Fiore, E., Armstrong, T. P., Holland, G., Lanzerotti, L. J. (1998). Electron, proton, and alpha monitor on the advanced composition explorer spacecraft. *Space Science Reviews*, 86, 541–562. https://doi.org/10.1023/A:1005088115759.

[29] Zwickl, R. D., Doggett, K. A., Sahm, S., Barrett, W. P., Grubb, R. N., Detman, T. R., Raben, V. J., Smith, C. W., Riley, P., Gold, R. E., Mewaldt, R. A., Maruyama, T. (1998). The NOAA real-time solar-wind (RTSW) system using ACE data. *Space Science Reviews*, 86, 633–648. https://doi.org/10.1023/A:1005044300738.

[30] Adriani, O., Barbarino, G. C., Bazilevskaya, G. A., Bellotti, R., Boezio, M., Bogomolov, E. A., Bongi, M., Bonvicini, V., Bottai, S., Bruno, A. (2017). Ten years of PAMELA in space. *Rivista del Nuovo Cimento*, 10, 473–522. https://doi.org/10.1393/ncr/i2017-10140-x.

[31] Xiaochao, Y., *et al*, (2024). A Multi-Satellite Survey Scheme for Addressing Open Questions on the Earth's Outer Radiation Belt Dynamics, Advances in Space Research, ISSN 0273-1177, doi:10.1016/j.asr.2024.08.008.

[32] Sawyer, D.M., Vette, J.I., (1976) AP-8 Trapped Proton Environment for Solar Maximum and Solar Minimum, NSSDC/WDC-A-R&S 76–06, NASA Goddard Space Flight Center, Greenbelt, Maryland.

[33] Reeves *et al.* (2003), "Acceleration and Loss of Relativistic Electrons during Geomagnetic Storms" (Geophysical Research Letters).

[34] Bartoloni, A., Strigari, L. (2021). Can high energy particle detectors be used for improving risk models in space radiobiology? *Proceedings of the Global Space Exploration Conference 2021 (GLEX2021)*, June 2021.

[35] Slaba, T., Blatting, S. (2014). GRC environmental model I: Sensitivity analysis for GCR environments. *Space Weather*, 12(4), 217–224.

[36] Bartoloni, A., Guracho, A. N., Della Gala, G., *et al.* (2022). Dose-effects models for space radiobiology: An overview on dose-effect relationship. *Proceedings of the International Astronautical Congress, IAC*, Volume 2022-September 2022, 73rd International Astronautical Congress, ISSN 00741795.

[37] Bartoloni, A., Strolin, S., and Strigari, L. (2020). Radiobiology with the Alpha Magnetic Spectrometer (AMS02) experiment on the International Space Station. *Proceedings of RAD 8 — 8th International Conference on Radiation in Different Fields of Research*, June 2020.

[38] Bartoloni, A., Baffioni, F., Bisello, F., *et al.* (2024). Bridging the gap: Exploring AMS, astroparticle experiments and space radiobiology. *Proceedings of Science*, Volume 44921, March 2024, Article number 0902023, European Physical Society Conference on High Energy Physics, EPS-HEP 2023, ISSN 18248039.

[39] Bartoloni, A., Della Gala, G., Guracho, A. N., Paolani, G., Santoro, M., Strigari, L., Strolin, S. (2022). High energy physics astro particle experiments to improve the radiation health risk assessment for humans in space missions. *EPS-HEP2021*, 398, 106. https://doi.org/10.22323/1.398.0106.

[40] Corti, C., Whitman, K., Desai, R., Rankin, J., Strauss, D. T., Nitta, N., Turner, D., Chen, T. Y. (2022). Galactic cosmic rays and solar energetic particles in cis-lunar space: Need for contextual energetic particle measurements at Earth and supporting distributed observation. Preprint at Arxiv. https://doi.org/10.48550/arXiv.2209.03635.

[41] Dong, Y., Zhang, S., Ambrosi, G. on behalf of the HERD collaboration. (2019). Overall status of the high energy cosmic radiation detection facility onboard the future China's Space Station. *Proceedings of the 36th International Cosmic Ray Conference (ICRC2019)*. https://doi.org/10.22323/1.358.0062.

[42] Adriani, O., Altomare, C., Ambrosi, G., Azzarello, P., Barbato, F. C. T., Battiston, R., Baudouy, B., Bergmann, B., Berti, E., Bertucci, B., *et al.* (2022). Design of an antimatter large acceptance detector in orbit (ALADInO). *Instruments*, 6(2), 19. https://doi.org/10.3390/instruments 6020019.

[43] Schael, S., Atanasyan, A., Berdugo, J., Bretz, T., Czupalla, M., Dachwald, B., *et al.* (2019). AMS-100: The next generation magnetic spectrometer in space — An international science platform for physics and astrophysics at Lagrange Point 2. arXiv. https://doi.org/10.1016/j.nima.2019.162561.

Chapter 4

Monitoring Space Radiation

4.1 Introduction

In the previous chapter, we focused on measuring the physical characteristics that identify space radiation (SR), such as charge, atomic number, momentum, and kinetic energy, associated with individual charged particles and nuclei. This section explores the methods and instruments used to measure and quantify astronauts' SR exposure and calculate dosimetry quantities to ensure their safety during missions. These measurements are crucial for assessing astronauts' radiation exposure in space. Dosimetry involves quantifying absorbed dose, equivalent dose, and effective dose, which are key indicators of the potential biological effects of radiation on human tissues. Instruments such as dosimeters, tissue-equivalent proportional counters (TEPCs), and radiation area monitors (RAMs) provide real-time monitoring and accurate dose assessments, ensuring astronauts remain within safe exposure limits.

After reviewing the historical and future space missions dedicated to measuring SR, we highlight the potential complementarity between the instruments used in these missions and those discussed in the previous section. The interplay between scientific research instruments and dosimetry tools enhances our understanding of cosmic rays and strengthens efforts to protect human health in space. By distinguishing between the two main objectives of SR measurement, advancing fundamental scientific knowledge and ensuring astronaut safety, we can appreciate the comprehensive approach required to tackle the challenges of space exploration. This dual focus underscores the importance of precision and innovation in developing and deploying instruments to accurately measure and understand SR.

4.2 Basic Principles of Monitoring Space Radiation

Given the uniqueness of the SR environment and its impact on crewed space exploration, a specialized radiation protection approach is warranted for space explorers.

As general concepts, the classification of radiation measurement instruments used in space dosimetry based on their functionality and application is summarized as follows:

- *Passive detectors*:
  - *Individual dosimetry*: Measures radiation exposure to an individual using emulsion, thermoluminescent dosimeters (TLD), nuclear track detectors, internal biodosimetry, radioactivation of tissue, and optically stimulated luminescence.
  - *Area dosimetry*: Measures radiation in a specific area using similar methods, including emulsion, TLD, nuclear track detectors, and optically stimulated luminescence, as well as external biodosimetry, induced radioactivity, and other chemical or physical latency methods.
- *Active detectors*:
  - *Individual dosimetry*: Provides real-time or time-resolved data using devices such as pocket ionization chambers, portable silicon detectors, high dose or dose rate warning monitors, direct ion storage devices, TEPCs, scintillation counters, particle spectrometers, and moderated neutron counters.
  - *Area dosimetry*: Includes instruments similar to individual dosimetry but is used to measure radiation levels in a specific area.

The distinction between passive and active detectors lies in their approach to measuring and characterizing radiation, either through integrated data or real-time monitoring.

The rest of the chapter examines the typical space mission scenarios that characterize the past and experiences of human travel and permanence on low Earth orbit (LEO), particularly aboard the International

Space Station (ISS) and during the first missions in deep space. Then, it examines the space scenarios that will characterize the following decades, particularly beyond low Earth orbit (BLEO) travel and Moon and Mars human explorations and permanence.

4.3 Apollo Space Missions and the Precursor Gemini Flights

The Apollo missions marked a significant milestone in space exploration, particularly in SR measurements. The Apollo radiation measurements and dosimetry program was meticulously designed to address these needs, drawing on lessons learned from earlier missions, such as Gemini, which, in the early 1960s, provided the first crewed space missions as precursor flights to Apollo. In the Gemini flights, which did not reach deep space, the only SR that spacecraft occupants received was during brief passes through the South Atlantic Anomaly (SAA) region [1].

4.3.1 *Radiation exposure during Gemini missions*

The experience gained from the Gemini missions provided valuable insights into radiation exposure in space. In that case, due to the low levels of radiation, dosimetry monitoring would only serve medical record purposes, so a passive system was developed and flown on the Gemini missions. It consists of a badge with several components, measuring 1.7 × 0.25 inches and containing 500 mg of lithium fluoride TLD, nuclear emulsions, and several standard films sensitive to electrons, gamma rays, and neutrons. The components were sandwiched between 0.011 in. polyvinyl-chloride films to make a soft, flexible package that the astronauts could wear in different locations on their bodies (chest, helmet, and thigh). The TLD can be read immediately after the flight. One of the TLD system's key components was emulsions, specifically 200-micron-thick layers of Ilford G-5 and K-2, which allowed for separate radiation dose estimates from different components. These emulsions were complemented by Kodak films, including a type 2 double-component pair and an NTA neutron monitoring film, providing a densitometric readout capability. This system ensured that the films could still offer reliable data, even if the emulsions were overexposed.

The data revealed that most radiation doses were relatively low, except for the Gemini V, VII, and X flights. Notably, Gemini V lasted for eight days and reached apogees of 200 nautical miles, while Gemini VII was a 14-day mission in a 160 nautical mile circular orbit. These extended durations and Gemini X's higher altitude in the SAA contributed to higher radiation exposure. The Gemini XI mission reached an apogee of 750 nautical miles and was explicitly programmed to avoid the anomaly by attaining its highest altitude over Australia. This was done to protect a nuclear emulsion cosmic-ray experiment. The variations in doses recorded on the same flights were often attributed to local shielding by the space-craft structure, highlighting the importance of real-time dose readout capabilities during missions.

4.3.2 *Apollo dosimetry design and mission operation procedures*

Stringent requirements guided the design of Apollo's dosimetry system, ensuring accurate radiation measurements while minimizing the impact on spacecraft resources. After extensive study, specific guidelines were established to direct the design and operational procedures of dosimetry instruments during missions, representing a standard for future space missions.

They can be summarized in the following points and considerations:

- The Command Service Module (CSM), the starship used in the Apollo missions, has a sufficiently robust structure to protect the crew adequately from even severe solar particle events.
- The most vulnerable part of the mission, concerning radiation, is during lunar operations when two crew members are shielded only by the lunar module (LM) and their space suits. Theoretically, the astronaut could receive skin doses of several thousand rads in these circumstances, based on different possible solar flare events that have been measured in the past. A ground-based warning system will monitor the Sun at radio and optical frequencies to preclude this possibility. Once word is received that an event has occurred, the crew can return to the CSM before the intense portion of the particles arrives.
- Both integrated dose and dose rate should be monitored, primarily for surface dose. This became evident after studies involving solar event spectral data and CSM shielding geometry showed that doses that produce skin erythema are probably not biologically significant at blood-forming organ (BFO) depths. The doses are well below the threshold for nausea in most of the general population. Should the

BFO dose be required, a particle spectrometer on the Apollo vehicle telemeters back real-time data that can be used to calculate the depth dose.

- Since no single location within the vehicle represents the varied shielding conditions the astronaut experiences, it is crucial that each crew member carries the integrating dosimeter, at least during a particle event. Because only the surface dose will be monitored, the dosimeter can be made small, light, and self-contained.

4.3.2.1 *Apollo missions' radiation monitoring equipment*

Various specialized dosimeters were developed to monitor radiation exposure during the Apollo missions, each tailored to specific monitoring needs and operational contexts. These include personal radiation dosimeters (PRDs), a radiation survey meter (RSM), passive dosimeters, and a nuclear particle detection system (NPDS).

The PRD was a compact, tissue-equivalent ionization chamber designed to monitor astronauts' radiation exposure continuously. Its structure incorporated necessary electronics, a self-contained battery power supply, and a signal readout. The ionization chamber's wall thickness was equivalent to 1 mm of tissue plus the astronaut's apparel, ensuring accurate surface dose measurement. The battery enabled up to 1,200 h of continuous operation. The PRD had a measurement range of 0–1,000 rad, with increments of 0.01 rad per pulse and a maximum pulse rate of three pulses per second, corresponding to a dose rate of 108 rad/h. The ionization chamber had an active volume of 7.1 cm^3, with a sensitivity of approximately 10^{-12} amp/rad/h. To minimize leakage, a high-impedance input circuit using a field effect transistor (FET) for amplification was enclosed within an evacuated can. Readouts were provided via a five-digit electromechanical register. The PRD was positioned on the left thigh in a pocket of the Apollo suit, ensuring continuous measurement of the astronaut's exposure during spaceflight and lunar surface operations.

The RSM was a portable, handheld device used to measure dose rates during spaceflight and lunar operations. The RSM featured a 10 cm^3 tissue-equivalent ionization chamber, solid-state circuitry, and a self-contained power supply. Constructed of tissue-equivalent plastic (TEP) filled with ethylene gas at 1 atm, the detector had a sensitivity of approximately 10^{-12} amp/rad/h. As in the PRD, a FET was employed in the amplification stage, replacing the conventional electrometer tube. The wall thickness ensured that dose measurements corresponded closely to surface doses. The RSM provided readings across four ranges: 0–0.1, 0–1, 0–10, and 0–100 rad/h, with a spring-loaded snubbing switch allowing the meter to lock at any specific reading. Designed for operational flexibility, the RSM could function continuously for up to 800 h or intermittently for 1,200 h without requiring a battery replacement. During missions, the RSM was carried aboard the command module and transferred to the LM for measurements on the Moon.

Passive dosimeters, similar to those used in the Gemini missions, were employed to provide additional radiation exposure data. These dosimeters were placed on various parts of the astronaut's body, including the right chest, left thigh, right ankle, and helmet. By capturing localized exposure measurements, passive dosimeters contributed to a more comprehensive understanding of the radiation environment and its effects on the human body during the mission.

The NPDS, mounted on the service module, served as a proton-alpha spectrometer for real-time monitoring and analysis of the radiation environment. The NPDS consisted of three solid-state detectors interspersed with absorbers, providing energy measurements across four increments for protons (15–150 MeV) and three for alpha particles (44–300 MeV). Despite its compact design, measuring 83 in^3, weighing 5.5 lbs., and consuming 1.60 watts of spacecraft power, the system delivered vital spectral and flux data to ground control via telemetry. These data were utilized for several purposes, including confirming the arrival of solar particles near the command module, calculating radiation doses within the module, and assisting in dose projections for mission planning. The NPDS played a critical role during solar flare events. Initial detection of high-energy particles confirmed the directionality of solar events, while subsequent lower-energy particles allowed for detailed spectral measurements of alpha and proton components. These real-time spectra enabled accurate dose calculations at various points in the command module, providing essential input for decisions about mission duration or potential aborts.

The advancements in radiation monitoring developed for the Gemini and Apollo missions established a foundational framework for all subsequent crewed space missions in LEO. These techniques and technologies remain essential for ensuring astronaut safety during future crewed explorations and extended stays BLEO.

4.4 The International Space Station on the LEO

The ISS is a science laboratory in LEO. Its orbit is at an altitude of about 400 km, placing it within the inner Van Allen belt (IVAB). Due to its specific orbit, the ISS can fly through the SAA, where it experiences heightened radiation exposure. For an estimation of the time spent in the SAA, considering that the ISS orbits the Earth approximately 16 times a day and travels at an average speed of about 27,600 km/h, it takes approximately 90 minutes to complete one orbit. The SAA region is about 5,000 km wide in the direction of the ISS's orbit, so the ISS would take around 12 minutes to cross it.

Given that the ISS might cross the SAA 3–4 times during its daily orbits, it could spend around 30–50 minutes per day in the SAA. This exposure is a significant concern for human health; nevertheless, several radiation protection mechanisms are implemented, such as physical shielding, water shielding, and careful orbital positioning. These strategies are crucial for safeguarding astronauts from the harmful effects of SR. Further, some monitoring systems equip the ISS for medical purposes and to characterize the different ISS locations.

4.4.1 *Radiation monitoring systems on the ISS*

The ISS has several monitoring systems that continuously measure radiation levels inside the station. These data help mission control assess the radiation exposure risk to astronauts and take appropriate action. In particular, the Space Radiation Analysis Group (SRAG) at the Johnson Space Center has provided continuous crew-worn radiation monitoring for each ISS crew member since the inception of the space station. These systems can be categorized based on their functions:

- *Active dosimetry systems*: Actively measure the real-time radiation dose received by astronauts and onboard equipment.
- *Passive dosimetry systems*: Designed to measure accumulated radiation exposure over a specific period. They are usually analyzed post-flight.

The following describes a series of instruments used to characterize the ISS radiation environment, whose technologies will also be used for dosimetry monitoring in future space missions.

4.4.1.1 *CAD, RAD, REM, and RMS dosimetry systems*

The Crew Active Dosimeter (CAD) [2] is a portable dosimetry system based on the Instadose® direct-ion storage (DIS) technology. The DIS consists of a floating-gate MOSFET surrounded by a conductive wall, creating an ionization chamber. The ionization-induced charge is collected on the MOSFET gate. It can be read as a non-destructive electrical signal worn by astronauts, measuring the dose of ionizing radiation in real time. This dosimeter is crucial for monitoring individual astronaut exposure during extravehicular activities (EVAs), where the crew is more vulnerable to galactic cosmic rays (GCRs) and solar energetic particles (SEPs).

The Radiation Assessment Detector (RAD) [3] is another essential tool for measuring radiation aboard the ISS, initially developed for the Mars Science Laboratory mission. It includes a charged particle telescope, which is currently deployed in the US lab and tied into the "Caution & Warning" system to alert the crew in case of an energetic solar particle event. The instrument consists of a charged particle detector (CPD) and a fast neutron detector (FND). The CPD consists of multiple solid-state detectors. Each of these detectors is 300 μm thick and operates at complete depletion. For dosimetry purposes, a dedicated trigger is defined so that all energy deposits above a threshold are recorded, added to a running total, stored, and cleared once per minute.

The threshold is set to ensure that even the smallest valid charged-particle deposits from typically incident minimum-ionizing particles are well above the threshold.

The most probable energy deposit for such particles is about 75 keV, and the threshold is set to 50 keV, which is low enough to allow the recording of all contributions and high enough to be above the noise level. At the other end of the range, the response remains linear for energy deposits up to 600 MeV. The per-minute running totals of deposited energy are converted to dose rates onboard the ISS using known detector mass and acquisition live-time and are output as part of the ISS's cyclic data stream. The raw data are also stored for subsequent telemetry to the ground, where identical analyses are performed to convert energy deposited into dose rates. During SAA passes, the dose rates are high, and not all events can be processed onboard; the counts of processed and unprocessed events are stored (per minute) and are used to correct the measured dose rates when necessary. During high dose rates caused by SEP events, RAD is tied to the ISS's caution and warning alert system. The RAD alarm will be triggered based on in-flight dose rate calculations similar to those performed during SAA pass transits. The dose measured as described is in silicon, which is converted to the approximate dose in water by applying a factor of 1.23.

The Radiation Environment Monitor (REM) [4], operated by the ESA, offers additional monitoring by detecting high-energy particles from cosmic rays and solar events. REM is installed at fixed locations on the ISS and provides data on radiation flux and particle energy spectra, particularly for electrons, protons, and heavy ions. These measurements help characterize the radiation environment and inform risk assessments for long-term missions. The Radiation Environment Monitor 2 (REM2) units are Timepix-based hybrid pixel detectors with Advacam Minipix readout using 500 μm thick sensor layers. There are currently seven REM2 units used as operational area monitors aboard the ISS. The REM2 unit is a small, compact instrument, approximately the size of a USB stick, weighing 56 g. Each REM2 is read and powered via a space station computer (SSC) USB connection. The Timepix detector comprises a 500 μm thick pixelated silicon sensor attached to the underlying Timepix application-specific integrated circuit (ASIC) via flip-chip solder bump bonding. The Medipix2 collaboration at CERN designs the Timepix ASIC. Its salient feature is that each semiconductor pixel is attached to a complete pulse processing chain, including a preamplifier, shaper, threshold discriminator, and Wilkinson-type analog-to-digital converter (ADC), all of which fit into the footprint of the overlying pixel. The Timepix detectors consist of 256×256 pixels of pitch 55 μm, for a total area of 2 cm^2, with 65,536 active pixels. Each Timepix pixel is individually calibrated to the deposited energy, with a minimum threshold of 5 keV. The effect of this whole matrix is that traversing particles create characteristic tracks in the sensor, much like a solid-state nuclear emulsion. By processing the tracks' energy deposit and chord lengths, the per-particle LET can be extracted, while the dose is measured by simply summing the total energy imparted to the sensor. The water dose is calculated by multiplying the dose in silicon by a standard conversion factor of 1.24.

The Radiation Monitoring System (RMS) [5] has continuously operated in various configurations since the launch of the Zvezda module of the ISS. It consists of seven units: the R-16 dosimeter, four DB-8 dosimeters, and utility and data collection units. It has been operating for more than 22 years, collecting the onboard daily dose rate associated with changes in ISS altitude and variations in GCRs over an 11-year cycle.

4.4.1.2 *DOSIS 3D and DOSTEL*

An example of a complex system, combining both active and passive dosimetry, is represented by the Dosimetry Telescope (DOSTEL) and the Dose Distribution Inside the ISS in 3D (DOSIS 3D) [6] instruments, which are used for radiation monitoring on the ISS. DOSTEL is an active dosimeter using silicon-based solid-state detection technology to measure absorbed dose and dose rate, providing directional information about the radiation field. It consists of a pair of silicon detectors arranged in a telescope configuration. It measures the energy deposition of charged particles and distinguishes between different types of radiation, such as protons, heavy ions, and electrons. It is typically used as a standalone device, often deployed in the Columbus module of the ISS for specific radiation monitoring tasks, and is part of the DOSIS 3D experiment.

In comparison, DOSIS 3D is a comprehensive radiation monitoring experiment that includes multiple DOSTEL units and other dosimetry instruments (such as passive dosimeters), aiming to map the spatial distribution of the radiation field inside the ISS in three dimensions over time. It measures the absorbed dose, dose rate, and directional information. It also examines the broader spatial distribution and temporal variations, providing as output spatial-temporal radiation dose distribution across multiple locations by using DOSTEL's detailed data from a single measurement point.

4.4.1.3 *ALTEA and LIDAL*

Similarly, the Anomalous Long-Term Effects on Astronauts (ALTEA) active dosimeter and particle detector and the Light Ion Detector for ALTEA (LIDAL) integrate the ALTEA capability specifically for light ions [7]. ALTEA, deployed in the US lab "Destiny," is based on a sophisticated array of silicon detectors designed to measure the flux, energy, and charge of incoming particles. It can detect various radiation types, including cosmic rays and solar particles. A TOF system provides additional information on the speed and mass of the particles, allowing for detailed characterization of the radiation environment. The LIDAL extension has a silicon detector array equipped with an enhanced TOF system and additional particle detection capabilities. It aims to augment the capabilities of ALTEA by providing enhanced detection and characterization of light ions (such as protons and helium nuclei), as well as improved sensitivity to heavy ions. It is specifically designed to enhance the detection of light ions, which are more prevalent in SR and contribute significantly to the radiation dose experienced by astronauts. LIDAL can measure particles' energy, charge, and velocity, allowing for a more detailed analysis of the radiation environment.

Table 4.1 represents the characteristics of the DOSIS3D & DOSTEL and ALTEA & LIDAL active dosimeter systems present on the ISS.

4.4.1.4 *SpaceTED*

The Space Technology Exposure for Dosimetry (SpaceTED) dosimeter is a more recent technology, operating on the ISS since 2018. It is designed to evaluate radiation exposure in space environments and has been tested on board the ISS. It uses novel dosimetry techniques, often leveraging advanced materials and sensors, to provide highly accurate real-time radiation dose measurements. It uses tissue-equivalent detectors [8] to simulate how human tissue would respond to radiation, making it especially useful for understanding radiobiological risks.

Table 4.1. Comparison of the measurement characteristics of DOSTEL & DOSIS 3D versus ALTEA & LIDAL on the ISS, highlighting the types of particles they measure and their corresponding energy ranges.

System	Measured Particles	Energy Range
DOSIS3D & DOSTEL [6]	Protons	10 MeV – 1 GeV
	Electrons	0.5 MeV – 10 MeV
	Light ions (e.g., He)	50 MeV/nucleon – 1 GeV/nucleon
	Heavy ions (e.g., Fe)	100 MeV/nucleon – several GeV/nucleon
ALTEA & LIDAL [7]	Protons	30 MeV – 700 MeV (ALTEA) 10 MeV – 1 GeV (LIDAL)
	Electrons	0.5 MeV – 10 MeV
	Light ions (e.g., He)	50 MeV/nucleon – 1 GeV/nucleon
	Heavy ions (e.g., Fe)	100 MeV/nucleon – 1 GeV/nucleon (ALTEA) 100 MeV/nucleon – several GeV/nucleon (LIDAL)

4.4.1.5 *TLDs and PNTDs*

Examples of passive dosimeters commonly used aboard the ISS include TLDs [8] and plastic nuclear track detectors (PNTDs) [9].

TLDs are strategically placed at various locations around the ISS. These small, lightweight devices absorb ionizing radiation over time, storing energy in their crystal lattice structures. TLDs used on the ISS are typically made from materials such as lithium fluoride (LiF), calcium fluoride (CaF_2), or magnesium silicate (Mg_2SiO_4). Each of these materials has unique properties that make them sensitive to different types of radiation. For example, lithium fluoride (LiF) is widely used due to its sensitivity to a broad range of ionizing radiation and its tissue-equivalent properties. When the TLDs are returned to Earth, they are carefully analyzed by heating them to release the stored energy as light. The intensity of this light is directly proportional to the amount of radiation to which the TLDs were exposed, allowing for precise determination of the total absorbed dose. TLDs are particularly valued for their ability to measure a wide range of radiation doses with high sensitivity, making them essential for long-term dosimetry studies in the ISS's variable radiation environment.

PNTDs are another type of passive dosimeter used aboard the ISS, specifically designed to measure the flux and energy spectrum of heavy ions in space. These detectors are composed of polymer films that, when struck by high-energy heavy ions, produce microscopic tracks along the paths of the ions. PNTDs are typically made from polycarbonate or CR-39 (allyl diglycol carbonate), which are polymers capable of recording the passage of high-energy ions through the material. When high-energy heavy ions traverse the PNTD, they create latent tracks in the polymer matrix. After returning to Earth, these detectors undergo a chemical etching process that reveals these tracks, allowing for detailed microscopic analysis. These tracks' size, shape, and density provide information about the type, energy, and direction of the heavy ions. PNTDs are particularly effective at measuring high-charge (Z) particles, such as cosmic ray nuclei, which are a significant component of SR and pose considerable risks to astronauts and spacecraft systems.

TLDs and PNTDs are invaluable tools for understanding the complex radiation environment within the ISS. They provide essential data that complement the real-time monitoring performed by active dosimeters, contributing to the overall radiation protection strategy and risk assessment for crew members and experiments aboard the station.

Tables 4.2(a) and 4.2(b) provide an overview of the radiation dosimeters used on the ISS, categorized by dosimeter type, technology, measurement type, and location on the ISS.

Table 4.2. Summary of some radiation dosimetry systems used on the International Space Station (ISS) over time: (a) active systems and (b) passive systems.

(a)			
Name	Detector Technology	Measurement Type	Location
CAD [2]	Silicon	Absorbed & equivalent dose, dose rates	Mobile is also used in the EVA
RAD [3]	Silicon & plastic scintillator	Absorbed & equivalent dose, dose rates	Fixed or research-specific modules across the ISS
REM [4]	Silicon	Radiation environment monitoring	Columbus module
RMS [5]	Silicon	Radiation environment monitoring, protons, electrons	Module "Zvezda"
DOSTEL [6]	Silicon	Absorbed dose, equivalent dose, dose rates, daily dose	Columbus Module
ALTEA & LIDAL [7]	Silicon, time of flight	Heavy nuclei, protons, electrons, radiation environment monitoring, particle flux	US Lab "Destiny"
SpaceTED [8]	TEPC, silicon	Absorbed & equivalent dose, dose rates	Module "Kibo"

(Continued)

Table 4.2.   (*Continued*)

(b)

Name	Detector Technology	Measurement Type	Location
OSLD (Optically Stimulated Luminescence Detector) [13]	TLD	Total Ionizing Dose (TID)	Various (mobile)
PADLES (Passive Dosimeters for Lifescience Experiments in Space) [14]	TLD, track-etch	TID, particle flux	Japanese module
DOBIES (Thermoluminescence Detector) [15]	TL & OSL	TID	Various (mobile)
PILLE [12]	TLD	TID	Various (mobile) & EVA
R-16 [6]	TLD	TID	Module "Zvezda"

4.4.2 *Characterization of the external radiation on the ISS*

These systems are crucial for monitoring the overall radiation environment around the ISS. They are designed to assess the radiation hazards in the space surrounding the station, including areas outside the ISS. By providing data on the external radiation environment, these systems help us understand the intensity and nature of radiation the ISS encounters as it orbits Earth. This information is essential for evaluating potential risks to astronauts and for developing effective radiation protection strategies. One example is the R3DR2 instrument on the ESA EXPOSE-R2 platform, located outside the Russian Zvezda module of the ISS. Data were collected from 24 October 2014 to 11 January 2016 at an average altitude of 415 km and a 51.6° orbital inclination [11].

The findings, summarized in Table 4.3, were related to various components of SR and their relevance to the expected doses. The primary

Table 4.3. Summary of radiation dose rates and TID from various sources in space [11].

Radiation Source	Average Daily Dose Rate (μGy)	Minimum Daily Dose Rate (μGy)	Maximum Daily Dose Rate (μGy)
Galactic Cosmic Rays (GCRs)	71.6	71.2	102
Protons (South Atlantic Anomaly, SAA)	567	426	844
Relativistic Electrons/ Bremsstrahlung (Outer VAB)	278	0.64	2,962
Solar Energetic Particle (SEP) Events	Peak hourly: >5,000	–	2,848

function of these detectors is to identify the different components of SR and to refine the technique for doing so. Overall, they provided a detailed assessment of the radiation environment on the ISS, highlighting the variability of radiation exposure and the impact of different radiation sources. Table 4.3 includes the average, minimum, and maximum daily dose rates for GCRs, protons in the SAA, relativistic electrons/bremsstrahlung in the OVAB, and peak dose rates observed during SEP events.

Another essential activity regarding the external field of radiation concerns the astronauts' EVAs, during which the only protection they have is their space suits. Usually, passive dosimeters are used for this kind of monitoring [10]. In this context, the astroparticle experiments described in Section 3.2 also play crucial roles, particularly AMS-02, which has been installed and operational on the ISS since 2011.

4.4.3 *Dosimetry on the ISS for biological payloads*

Biological payloads assess the biological impact of radiation on living tissues, particularly to study how SR affects human health. To secure the scientific findings of such a payload, every location must be monitored and characterized accurately. The absorbed dose to biological payloads on the ISS depends on various factors, including the date and duration of exposure and the payload's location on the station. Several specific factors contribute to variations in the absorbed dose rate in orbit:

1. *Altitude*: The ISS orbits at altitudes ranging from 330 to 435 km.
2. *Solar Cycle*: The point within the 11-year solar cycle when the sample is exposed influences radiation levels, with solar activity affecting the intensity of GCRs.
3. *Local Shielding Distribution*: The most critical factor is the local shielding distribution around the biological sample. This varies depending on the module in which the sample resides and its location.

In addition to these factors, the measured absorbed dose can vary based on the dosimetry instruments used. Radiation monitoring on the ISS is not standardized across modules, as each international partner uses its own suite of instruments with differing technologies, sensitivities, and analysis methodologies. There are differences in criteria for distinguishing between lightly ionizing particles (mainly protons) and highly ionizing particles (mainly GCR nuclei).

The accuracy and precision of these measurements vary among instruments, which is reflected in the precision and errors presented in the tabulated data. For detailed information on how the quoted doses were obtained, it is essential to consult the cited literature for each payload and dosimeter.

The dose reported in the metadata corresponds to the dose recorded by the radiation detector closest to the sample. In most cases, that detector is located in the same ISS module. Suppose data from a detector in the same module are unavailable. In that case, the dose is extrapolated using data from periods when data from multiple modules, including the one containing the payload, are available.

The following lists the different locations on board the ISS that are characterized continuously:

- *COL*: "Columbus" (European Space Agency, ESA).
- *JPM*: Japanese Pressurized Module, part of the Japanese Experiment Module "Kibo" (Japanese Aerospace Exploration Agency, JAXA).
- *LAB*: US Lab "Destiny" (NASA).
- *NOD1, NOD2, and NOD3*: Nodes connecting ISS modules:
    - *Node 1 ("Unity")*: connects the US and Russian segments;
    - *Node 2 ("Harmony")*: connects the US, ESA, and Japanese segments;
    - *Node 3 ("Tranquility")*: attached to the port side of Node 1.
- *SMP*: Russian Service Module "Zvezda" (Roscosmos).

4.5 Radiation Characterization for BLEO Moon and Mars Travel and Exploration

Human exploration of the BLEO is on the agenda since few humans have ventured there. For a short-term permanence period, the next decade's exploration activities must be complemented by significant efforts to develop technologies for safe exploration. First, an intense characterization of the SR fields will be needed to support this development. The following four critical areas of interest will be taken into consideration.

4.5.1 *LEO and Van Allen belts characterization*

The Van Allen belts (VABs), consisting of energetic particles trapped by Earth's magnetic field, pose a significant radiation hazard for spacecraft traveling BLEO. Accurate modeling of this radiation environment is crucial for designing effective spacecraft shielding and ensuring the safety of astronauts and technology. Data from missions like NASA's Van Allen Probes have provided invaluable insights into the intensity and variability of radiation in these regions, showing that the environment within the VABs can change dramatically over time. These changes, driven by solar activity and geomagnetic conditions, underscore the need for advanced, dynamic models to predict these variations and mitigate associated risks for future missions [16].

Legacy radiation models [17], such as AP8-MIN and AP8-MAX, have provided foundational insights but are inherently static. AP8, for instance, is based on data collected in the 1960s and 1970s, and although it differentiates between solar maximum and minimum conditions, it only offers averaged non-real-time predictions. AP8's limitations become apparent when addressing radiation conditions in the dynamic space environment. It is primarily used for general mission analysis, including those of the ISS and Space Shuttle operations. However, it does not adjust for short-term variations in space weather, making it insufficient for modern mission requirements.

In contrast, the AP9 [18] model represents a significant step forward, providing dynamic radiation predictions. To accurately model the trapped proton environment, AP9 incorporates many modern satellite datasets, including those from the Van Allen Probes and other missions. It uses Monte Carlo simulations to provide statistical estimates of radiation exposure, making it highly valuable for spacecraft operating in high-energy environments such as the medium Earth orbit (MEO) and highly elliptical orbits (HEO). AP9 also adjusts for variations due to solar activity and geomagnetic conditions, offering more precise, real-time predictive capabilities.

While static models such as AP8 focus on adiabatic processes and collisional loss mechanisms, they fail to account for the dynamic changes in the radiation environment. These include influences from solar activity, geomagnetic storms, and particle transport processes, such as inward radial diffusion and local acceleration by Whistler-mode chorus waves. In the OVAB, electron phase space density is significantly altered by such interactions, with ultralow-frequency (ULF) waves playing a key role in redistributing particles. Static models cannot capture these time-dependent processes.

Future modeling efforts must, therefore, emphasize dynamic modeling to better understand and predict the complex behaviors of the VABs. Unlike their static predecessors, dynamic models can incorporate real-time solar and geomagnetic data, accounting for processes such as electron injections from substorms, inward radial diffusion, and local heating effects from waves. Furthermore, loss mechanisms, such as magnetopause compression caused by solar wind pressure and pitch-angle scattering from electromagnetic ion cyclotron (EMIC) waves, are crucial for predicting electron precipitation into the upper atmosphere. Combining these dynamic factors with new tools such as machine learning offers the potential to enhance the accuracy of predictions, leveraging data from multiple satellites and providing a comprehensive understanding of Earth's radiation environment.

Several space missions have played a pivotal role in validating radiation belt models and improving our understanding of the dynamic processes within the VABs [19].

The Combined Release and Radiation Effects Satellite (CRRES), launched in 1990, observed dramatic variations in the radiation belts, leading to a new era of analysis. The Akebono satellite, launched in 1989, focused on auroral electrons and the polar ionosphere. The Solar Anomalous and Magnetospheric Particle Explorer (SAMPEX), launched in 1992, conducted long-term observations over a solar cycle, indicating that interplanetary conditions influence the dynamics of the outer radiation belt. More recently, the Van Allen Probes, launched in 2012, have provided a unique understanding of Earth's radiation belts. These probes measured particles across a broad energy range and significantly advanced our understanding of the outer radiation belts, including the discovery of an unexpected third-ring structure.

Additionally, various operational satellites, such as GOES, POES, Fengyun, and GPS, have continuously monitored the radiation belt to provide *in situ* information. Long-term observations from these satellites offer an alternative perspective on Earth's radiation belts.

More recently, the CROPIS mission, launched in 2018 with RAMIS detectors on board, continues to provide vital data on absorbed dose and dose rates at an altitude of around 500 km, contributing to a more complete understanding of the global radiation environment [20], and in 2024, as part of the Geostationary Operational Environmental Satellite (GOES) space missions, the Energetic Heavy Ion Sensor (EHIS) within the Space Environment In-Situ Suite (SEISS) represents a significant advancement in space weather monitoring. EHIS can operate without saturation during extreme SEP events, reaching up to three times the intensity of the most energetic SEP known. This capability is critical for improving our understanding of radiation dynamics in the VABs and beyond. The SEISS suite also contributes to early warnings of high flux events, helping to mitigate potential damage to spacecraft and satellites.

While significant progress has been made in understanding and modeling the VABs, challenges remain, particularly in coupling radiation belt models with global geospace models for more accurate space weather forecasting. To address these challenges, future investigations must integrate multi-satellite data with ground-based observations and cutting-edge dynamic modeling tools. Incorporating nonlinear wave–particle interactions and machine-learning approaches will be critical for developing the next generation of radiation belt models. These advancements will ensure that radiation belt modeling continues to play a vital role in protecting technology and human space exploration, much like how meteorological models predict weather conditions on Earth today.

4.5.2 *Moon missions*

Several factors, including the lack of a significant atmosphere and magnetic field, influence radiation exposure on the Moon. This leaves lunar missions vulnerable to higher levels of GCRs and SEPs. Historical missions, such as Apollo, provided foundational data on the lunar radiation environment; however, updated models are now required to address current and future mission parameters. Recent studies emphasize the importance of advanced radiation shielding and real-time monitoring in protecting astronauts during extended stays on the lunar surface. Proper characterization of lunar radiation will inform the development of protective measures and mission planning to ensure crew safety.

One of the first space missions to measure the lunar orbit's radiation environment was the Cosmic Ray Telescope for the Effects of Radiation (CRaTER) aboard the Lunar Reconnaissance Orbiter (LRO) [21]. CRaTER measures ionizing energy loss in silicon solid-state detectors and TEP [22], a synthetic analog of human tissue, to assess the effects of SEPs and GCRs. It provides direct measurements of the linear energy transfer (LET) spectrum, focusing on ions with energies above 10 MeV, and helps refine models that predict the biological effects of ionizing radiation. This also provides valuable insights for radiation protection in human exploration and deep-space electronics.

In addition to human exploration goals, CRaTER provides insights into the spatial and temporal variability of the SEP and GCR populations and their interactions with the lunar surface. Another critical aspect is radiation on the Moon's surface, including components from space and albedo radiation produced by backscattering phenomena when SR interacts with the lunar regolith.

The Chang'e 4 mission, which successfully landed on the far side of the Moon in January 2019, included the Lunar Lander Neutron and Dosimetry (LND) [23] experiment, aimed at characterizing the radiation environment. The LND instrument measures neutron and gamma radiation, providing essential dosimetry data to assess the radiation levels that astronauts may face during extended lunar stays. Results from this mission show that GCRs significantly contribute to radiation dose, with measurements showing a dose rate of approximately 60 μS per hour on the lunar surface, which is about 2.6 times higher than what astronauts experience aboard the ISS.

As an example of future technologies, we can consider the systems used in the BioSentinel [24] and the Astrobotic Peregrine lander missions, which utilize the Timepix [25] technology developed by CERN. This advanced detector measures LET, quantifying the energy deposited by ionizing radiation per unit length, which is crucial for assessing radiation quality. One of the challenges of using Timepix in space is the extensive data processing required to accurately interpret measurements. While high-energy physics (HEP) experiments can handle such data efficiently, satellite missions face unique challenges managing the data volume. Timepix allows researchers to analyze the overall quality factor of the radiation spectrum rather than individual particles, enabling a broader analysis of the radiation environment, especially considering the fixed component of GCRs.

This new technology will be integrated with biological payloads, such as the Lunar Explorer Instrument for Space Biology Applications (LEIA) program [26], developed at NASA Ames Research Center. LEIA aims to:

- combine yeast genetics with metabolic modeling to determine cellular sensitivity to the lunar environment;
- evaluate synthetic biology-enabled production of antioxidant nutrients and proteins under lunar surface conditions;
- assess genetically engineered yeast for enhanced tolerance to the lunar environment;
- measure and characterize biologically relevant radiation on the lunar surface.

LEIA is scheduled to land in the South Pole region of the Moon in 2026 and will be the first biology payload delivered through NASA's Commercial Lunar Payload Services (CLPS) initiative. These objectives enhance our understanding of how SR affects biological systems, particularly in the Moon's unique environment.

4.5.3 *Artemis program space missions*

NASA's Artemis program aims to return humans to the Moon and establish a sustainable presence [27]. Radiation characterization is a critical component of the Artemis missions, as the program will involve extended stays on the lunar surface and potential exploration of lunar resources. The Artemis missions will benefit from advanced radiation monitoring systems and updated radiation models incorporating data from previous missions and new insights from ongoing research. Effective radiation management strategies, including habitat shielding and monitoring of space weather conditions, will safeguard astronauts from GCRs and SEPs. NASA's Artemis program represents a reference point in this field. This program aims to explore the Moon's surface, send the first woman to the Moon, and obtain the necessary knowledge and experience required to send the first astronaut to Mars. Here is an overview of the planned missions from I to V:

- *Artemis I (2022)*: An uncrewed mission, completed in late 2022, that tested NASA's Space Launch System (SLS) rocket and the Orion spacecraft. It successfully orbited the Moon and returned to Earth to assess the system's readiness for crewed missions.
- *Artemis II (2026)*: The first crewed mission will take four astronauts on a 10-day journey around the Moon, performing a lunar flyby before returning to Earth. This will be crucial to evaluating systems with humans aboard for longer durations.
- *Artemis III (2027)*: This mission aims to land the next humans on the lunar surface, including the first woman. It will target the Moon's South Pole and use SpaceX's Starship as the lunar lander. Astronauts will stay for about a week and conduct exploration activities.
- *Artemis IV (2029)*: This mission will be critical for assembling NASA's Lunar Gateway, a space station orbiting the Moon to support long-duration missions. It will deliver the Gateway's Habitation and Logistics Outpost (HALO) module while astronauts may conduct further lunar explorations.
- *Artemis V (2030)*: This mission will expand lunar exploration and further develop infrastructure on the Moon's surface, supporting extended stays. The Lunar Gateway will be utilized for docking, and SpaceX will again provide the lunar lander.

In November 2022, NASA launched Artemis I. In the spacecraft, there were two female phantoms covered with many dosimeters (see Figs. 4.1(a) and 4.1(b)). The spacecraft was in space for several weeks and circled the Moon before returning to Earth in December 2022 (see Fig. 4.1(c)). Such a mission provides valuable dosimetry data for researchers [28, 29]. The characterization of the SR environment in the crew cabin was a key objective of Artemis I. Radiation was assessed using detectors at fixed locations in Orion throughout the Artemis I mission. Table 4.4(a) summarizes the different detectors used in the mission, and Table 4.4(b) contains some relevant results concerning the shielding characterization of different Orion capsule locations under different exposure conditions during the travel.

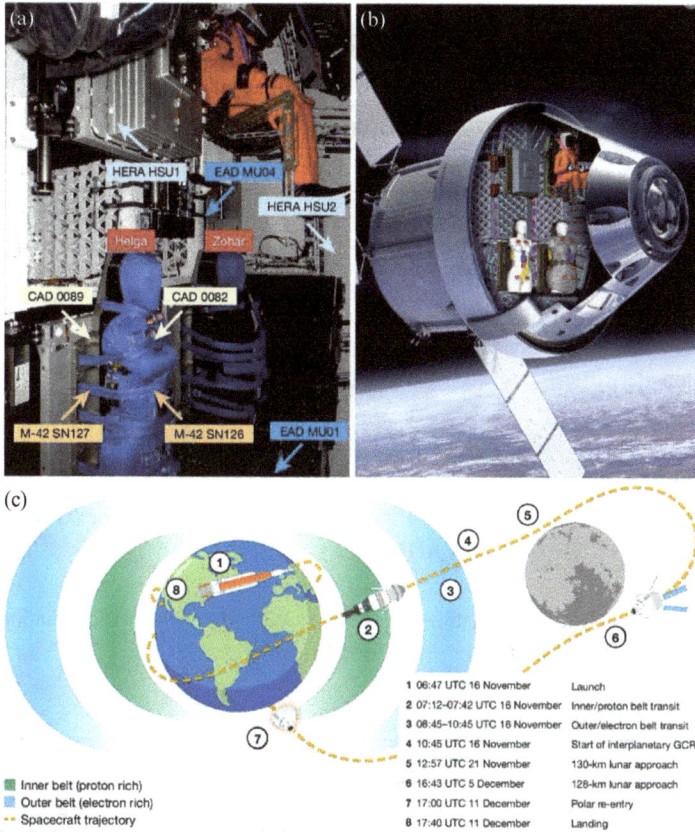

Fig. 4.1. (a) Radiation instrumentation and phantoms inside Orion. These consist of the NASA HERA system, the ESA EAD system, and the NASA CAD and DLR M-42 instruments. The HERA system and the EADs were hard-mounted at distinctly shielded Orion locations. CAD and M-42 were placed on the front and back surfaces (skin) and inside (organs) (M-42) of the MARE phantoms. (b) Placement of the instrumentation and hardware inside the Orion spacecraft. (c) The Orion flight profile concerning radiation for the NASA Artemis I mission. After launch at 06:47 UTC on 16 November 2022, Orion passed the inner (proton-dominated) and outer (electron-dominated) Earth radiation belts. Orion then ventured into interplanetary space dominated by GCRs. It passed the Moon twice on 21 November (first lunar flyby at 130 km) and on 5 December (second lunar flyby at 128 km). During these flybys, the Moon acts as a shield against GCRs. Orion re-entered Earth's atmosphere over the South Pole and landed in the Pacific Ocean close to San Diego, California, on 11 December 2022 at 17:40 UTC.

Source: George, S.P., Gaza, R., Matthiä, D. *et al.* Space radiation measurements during the Artemis I lunar mission. *Nature* 634, 48–52 (2024). CC BY 4.0 International (https://creativecommons.org/licenses/by/4.0/).

Table 4.4.   (a) An overview of the instruments on board the Artemis I space mission [29]. (b) Key findings on radiation dose rates in different locations aboard Orion, validating the effectiveness of its shielding against the inner proton belt and solar-particle events.

(a)			
Objective	**Instrument**	**Agency**	**Measurement**
Characterization of Space Radiation	Hybrid Electronic Radiation Assessor (HERA)	NASA	Ionizing energy, absorbed dose (Gy), dose equivalent (Sv)
Characterization of Space Radiation	ESA Active Dosimeter (EAD)	ESA	Ionizing energy, absorbed dose (Gy), dose equivalent (Sv)
Characterization of Space Radiation	M-42	DLR	Absorbed dose (Gy)
Characterization of Space Radiation	Crew Active Dosimeter (CAD)	NASA	Absorbed dose (Gy)
Organ Dose Estimation	Matroshka AstroRad Radiation Experiment (MARE)	NASA/DLR	Organ-specific radiation doses

(b)			
Measurement Location	**Instrument**	**Dose Rate (μGy/min)**	**Shielding Observation**
Most Shielded	M-42	69	This location experienced the lowest dose rate, demonstrating effective shielding in Orion's design against inner proton belt radiation.
Least Shielded	EAD	240	It is one of the least shielded areas, producing higher radiation exposure than the most shielded regions.
Crew Cabin	HERA	287	The highest dose rate recorded emphasizes that locations with less shielding, such as the crew cabin, are more susceptible to higher radiation exposure.

Table 4.4.   (*Continued*)

(b)

Measurement Location	Instrument	Dose Rate (μGy/min)	Shielding Observation
Storm Shelter	HERA	134	The "storm shelter" showed significantly lower radiation exposure, validating its design to protect the crew during solar-particle events and high-radiation environments.
Simulated Solar-Particle Event (Crew)	HERA	414	The October 1989 event in the crew area was simulated, and higher dose rates were shown due to the energetic spectrum of protons.
Simulated Solar-Particle Event (Shelter)	HERA	95	The storm shelter showed reduced radiation exposure during the simulated event, confirming its protective role.

4.5.4 *Mars space missions*

Mars missions present unique challenges due to the planet's thin atmosphere and lack of a global magnetic field, exposing astronauts to elevated radiation levels from GCRs and SEPs. Radiation characterization for Mars exploration involves understanding the long-term effects of SR on human health, as missions to Mars will involve prolonged exposure periods. Research efforts are focused on developing robust radiation protection strategies, including advanced spacecraft shielding, habitat design, and medical countermeasures. Modeling of Mars' radiation environment will rely on data from spacecraft missions, such as the Mars Science Laboratory and upcoming missions, to refine risk assessments and ensure the safety of crew members during their journey and stay on the Martian surface.

The Mars Radiation Environment Experiment (MARIE), aboard NASA's 2001 Mars Odyssey mission, was the first to measure the radiation environment in Mars orbit [30]. MARIE collected data on cosmic rays and SEPs, which are essential for understanding the radiation hazards that future astronauts might face on Mars. Unfortunately, it ceased operations in 2003 after a powerful solar flare disrupted the equipment. Despite this, MARIE's findings helped establish baselines for Martian radiation levels and laid the groundwork for later missions, such as the Mars Atmosphere and Volatile Evolution (MAVEN) space mission [31].

NASA's MAVEN mission, launched in 2013, continues to explore how the Martian atmosphere interacts with solar wind and radiation. It is the first mission devoted to studying the upper atmosphere of Mars and its relationship with solar radiation. MAVEN's data have been critical in showing how Mars lost much of its atmosphere to space, mainly through solar wind's influence. The MARIE and MAVEN space missions studied radiation in Mars orbit, while on the surface, NASA's MSL-RAD (Radiation Assessment Detector) has been collecting data since the Curiosity rover landed in 2012 [32]. MSL-RAD has been measuring radiation levels since 2013, covering an entire solar cycle. The daily radiation dose on Mars' surface ranges between 160 and 340 µGy, with SEPs potentially raising doses to 10,000 µGy. Pressure modulation affects radiation levels, as atmospheric pressure fluctuations affect radiation exposure. Additionally, MSL-RAD has been used to characterize the shielding capability of Martian regolith through indirect measurements, offering potential strategies for the future protection of human explorers.

4.6 Toward a Common Platform for Measurements and Monitoring

The next generation of Cosmic Ray Detectors (CRDs) presents an opportunity to advance our understanding of cosmic rays and high-energy particles and to integrate critical systems for space radiobiology and weather monitoring. Incorporating these functionalities from the outset can optimize space missions' scientific returns and operational efficiency, particularly those of BLEO. Given the high costs and lengthy development timelines of these sophisticated instruments, a collaborative approach across space agencies is essential to foster new ideas for a standardized instrumentation platform. The development timelines and costs for CRDs are significant. Historical projects such as AMS-02 required over a decade of development, while future projects such as AMS-100 and ALADInO are expected to follow similar timelines, with the added complexity of BLEO deployment. AMS-100, for example, aims to be 100 times more sensitive than AMS-02, necessitating advancements in technology and infrastructure that will require substantial investment. ALADInO, positioned at the L2 point, will also involve significant power, communication, and operational stability challenges, contributing to extended development timelines and increased costs.

Given these challenges, the push toward a common platform for CRDs that incorporates space radiobiology and space weather monitoring. This is both a scientific priority and a strategic imperative for supporting the complex systems modeling of a unified heliospheric space weather environment [33]. By unifying efforts across different space agencies and fostering international collaboration, we can ensure that future CRDs are more cost-effective, have shorter development timelines, and deliver broader scientific value. This integrated approach will be essential for the success of future space exploration missions, particularly those venturing into the deeper reaches of our solar system.

4.6.1 *Integrated space radiobiology and space weather systems*

As human space exploration extends to the Moon, Mars, and potentially beyond, understanding and mitigating the effects of SR on the human body become increasingly important. Space radiobiology, which studies the impact of cosmic rays and solar radiation on living organisms, can benefit significantly from data provided by CRDs. We can accurately measure radiation doses during space missions by developing CRDs with built-in passive and active dosimetry systems. These measurements are crucial for assessing the risks to astronauts and developing effective shielding and other protective measures.

Furthermore, data from CRDs can enhance space weather monitoring, which involves tracking solar activity and its effects on space environments. Understanding space weather is vital for protecting astronauts and spacecraft from harmful radiation. Incorporating space weather sensors into CRDs would enable real-time monitoring of solar particle events and cosmic ray flux, providing early warnings and allowing for timely countermeasures to be implemented.

Given these constraints, a collaborative approach is desirable and necessary among space agencies and research institutions. Collaboration can help distribute the financial burden, accelerate development timelines, and enhance scientific output. By pooling resources and expertise, agencies can ensure that CRDs are equipped with state-of-the-art instrumentation for multiple scientific purposes, including space radiobiology and space weather monitoring.

In fostering new ideas for a typical instrumentation platform, it is helpful to consider different architectural approaches. The following provides a few examples:

Satellite-style missions: This approach resembles the James Webb Space Telescope (JWST), a standalone observatory located at the Earth–Sun L2 Lagrange point. For CRDs, a similar setup could be employed, where the detector operates independently in a stable orbit, away from the influences of Earth's magnetosphere. This would be particularly suitable for missions like ALADInO, designed to operate at L2, providing a dedicated platform for cosmic ray detection and radiobiological experiments.

Platform-based missions: CRDs could be integrated into larger platforms like the ISS or future space stations designed for BLEO missions. This approach offers the advantage of combining multiple experiments on a single platform, allowing for shared resources and infrastructure. For example, the ISS currently hosts CRDs such as AMS-02 and CALET, along with various other scientific instruments, enabling multidisciplinary research in a controlled environment.

Networks of cooperative instruments: A more innovative approach could involve deploying a network of smaller, cooperative instruments, similar to SpaceX Starlink's satellite network telecommunications operation. In this model, multiple CRDs could be deployed across different orbits or locations to provide a comprehensive picture of cosmic ray activity and space weather conditions. This distributed network could offer redundancy, increase coverage, and allow real-time data sharing among nodes.

References

[1] Davis, W. G., Lill, J. C., Richmond, R. G., and Warren, C. S. (1968). Radiation dosimetry on the Gemini and Apollo missions. *Journal of Spacecraft and Rockets*, 5(2), 207–210. https://doi.org/10.2514/3.29217.

[2] Gaza, R., Johnson, A. S., Hayes, B., *et al.* (2023). The importance of time-resolved personal Dosimetry in space: The ISS crew active dosimeter. *Life Sciences in Space Research*, 39, 95–105. https://doi.org/10.1016/j.lssr.2023.08.004.

[3] Hassler, M. D., *et al.* (2012). The Radiation Assessment Detector (RAD) investigation. *Life Sciences in Space Research*, 170. https://doi.org/10.1007/s11214-012-9913-1.

[4] Stoffle, N., Pinsky, L., Kroupa, M., *et al.* (2015). Timepix-based radiation environment monitor measurements aboard the International Space Station. *Nuclear Instruments and Methods in Physics Research, Section A: Accelerators, Spectrometers, Detectors, and Associated Equipment*, 782. https://doi.org/10.1016/j.nima.2015.02.016.

[5] Benghin, V., Shurshakov, V., Osedlo, V., *et al.* (2023). Results of long-term radiation environment monitoring by the Russian RMS system on board Zvezda module of the ISS. *Life Sciences in Space Research*, 39, 3–13. https://doi.org/10.1016/j.lssr.2022.11.002.

[6] Berger, T., Burmeister, S., Matthiä, D., Przybyla, B., Reitz, G., *et al.* (2017). DOSIS & DOSIS 3D: Radiation measurements with the DOSTEL instruments onboard the Columbus Laboratory of the ISS in the years 2009–2016. *Journal of Space Weather and Space Climate*, 7, A8. https://doi.org/10.1051/swsc/2017005.

[7] Romoli, G., *et al.* (2023). LIDAL, a Time-of-flight radiation detector for the International Space Station: Description and ground calibration. *Sensors*, 23(7), 3559. https://doi.org/10.3390/s23073559.

[8] Hayes, B. M., Causey, O. I., Gersey, B. B., Benton, E. R. (2022). Active tissue equivalent dosimeter: A tissue equivalent proportional counter flown onboard the International Space Station. *Nuclear Instruments and Methods in Physics Research Section A*, 1028, 166389. https://doi.org/10.1016/j.nima.2022.166389.

[9] Benton, E. V., Richmond, R. G. (1986). Applications of nuclear track detectors in space radiation dosimetry. *International Journal of Radiation Applications and Instrumentation. Part D. Nuclear Tracks and Radiation Measurements*, 12(1–6), 505–508. https://doi.org/10.1016/1359-0189(86)90639-4.

[10] Rana, M. A. (2018). CR-39 nuclear track detector: An experimental guide. *Nuclear Instruments and Methods in Physics Research Section A*, 910, 121–126. https://doi.org/10.1016/j.nima.2018.08.077.

[11] Dachev, T. P., Bankov, N. G., Tomov, B. T., Matviichuk, Y. N., Dimitrov, P. G., Häder, D.-P., and Horneck, G. (2017). Overview of the ISS radiation environment observed during the ESA EXPOSE-R2 mission in 2014–2016. *Space Weather*, 15, 1475–1489. https://doi.org/10.1002/2016SW001580.

[12] Hirn, A., *et al.* (2024). Upgrade of the Hungarian PILLE ISS onboard thermolunminescent system for the dose assessment during extravehicular activities. *Radiation Measurements*, 177, 107255. https://doi.org/10.1016/j.radmeas.2024.107255.

[13] Yukihara, E. G., Sawakuchi, G. O., Guduru, S., McKeever, S. W. S., Gaza, R., Benton, E. R., Yasuda, N., Uchihori, Y., Kitamura, H. (2006).

Application of the optically stimulated luminescence (OSL) technique in space dosimetry. *Radiation Measurements*, 41(9–10), 1126–1135. https://doi.org/10.1016/j.radmeas.2006.05.027.

[14] Nagamatsu, A., Murakami, K., Kitajo, K., Shimada, K., Kumagai, H., Tawara, H. (2013). Area radiation monitoring on ISS increments 17 to 22 using PADLES in the Japanese Experiment Module Kibo. *Radiation Measurements*, 59, 84–93. https://doi.org/10.1016/j.radmeas.2013.05.008.

[15] Vanhavere, F., Genicot, J. L., O'Sullivan, D., Zhou, D., Spurný, F., Jadrníčková, I., Sawakuchi, G. O., Yukihara, E. G. (2008). DOsimetry of BIological EXperiments in SPace (DOBIES) with luminescence (OSL and TL) and track etch detectors. *Radiation Measurements*, 43(2–6), 694–697. https://doi.org/10.1016/j.radmeas.2007.12.002.

[16] Li, W., and Hudson, M. K. (2019). Earth's Van Allen radiation belts: From discovery to the Van Allen Probes era. *Journal of Geophysical Research: Space Physics*, 124, 8319–8351. https://doi.org/10.1029/2018JA02594.

[17] Sawyer, D. M., Vette, J. I. (1976). AP-8 Trapped proton environment for solar maximum and solar minimum. *NSSDC/WDC-A-R&S 76–06*, NASA Goddard Space Flight Center, Greenbelt, Maryland.

[18] Ginet, G. P., O'Brien, T. P., Huston, S. L., *et al.* (2013). AE9, AP9 and SPM: New models for specifying the trapped energetic particle and space plasma environment. *The Van Allen Probe Mission*, 579–615. https://doi.org/10.1007/978-1-4899-7433-4_18.

[19] Xiaochao, Y., *et al.* (2024). A multi-satellite survey scheme for addressing open questions on the Earth's outer radiation belt dynamics. *Advances in Space Research*. https://doi.org/10.1016/j.asr.2024.08.008.

[20] Meier, M. M., Berger, T., Jahn, T., *et al.* (2023). Impact of the South Atlantic Anomaly on radiation exposure at flight altitudes during solar minimum. *Scientific Reports*, 13, 9348. https://doi.org/10.1038/s41598-023-36190-5.

[21] Spence, H. E., Case, A. W., Golightly, M. J., *et al.* (2010). CRaTER: The cosmic ray telescope for the effects of radiation experiment on the lunar reconnaissance orbiter mission. *Space Science Reviews*, 150, 243–284. https://doi.org/10.1007/s11214-009-9584-8.

[22] Hayes, B. M., *et al.* (2022). Active tissue equivalent dosimeter: A tissue equivalent proportional counter flown onboard the International Space Station. *Nuclear Instruments and Methods in Physics Research Section A*, 1028, 166389.

[23] Wimmer-Schweingruber, R. F., Yu, J., Böttcher, S. I., *et al.* (2020). The lunar lander neutron and dosimetry (LND) experiment on Chang'E 4. *Space Science Reviews*, 216, 104. https://doi.org/10.1007/s11214-020-00725-3.

[24] Rahmanian, S., Slaba, T. C., Braby, L. A., *et al.* (2023). Galactic cosmic ray environment predictions for the NASA BioSentinel mission. *Life Sciences in Space Research*, 38, 19–28. https://doi.org/10.1016/j.lssr.2023.05.001.

[25] Bergmann, B., Gohl, S., Garvey, D., Jelínek, J., Smolyanskiy, P. (2024). Results and perspectives of timepix detectors in space—From radiation monitoring in low Earth orbit to astroparticle physics. *Instruments*, 8, 17. https://doi.org/10.3390/instruments8010017.

[26] Kanapskyte, A., Hawkins, E. M., *et al.* (2021). Space biology research and biosensor technologies: Past, present, and future. *Biosensors*, 11, 1–10. https://doi.org/10.3390/bios11020038.

[27] NASA. (n.d.). Artemis program. https://www.nasa.gov/specials/artemis/.

[28] NASA. (n.d.). Artemis 1. https://www.nasa.gov/artemis-1.

[29] George, S. P., Gaza, R., Matthiä, D., *et al.* (2024). Space radiation measurements during the Artemis I lunar mission. *Nature.* https://doi.org/10.1038/s41586-024-07927-7.

[30] Badhwar, G. D. (2004). Martian radiation environment experiment (Marie). In C. T. Russell (Ed.), *2001 Mars Odyssey*. Springer, Dordrecht. https://doi.org/10.1007/978-0-306-48600-5.

[31] Jakosky, B. M., Lin, R. P., Grebowsky, J. M., *et al.* (2015). The Mars atmosphere and volatile evolution (MAVEN) mission. *Space Science Reviews*, 195, 3–48. https://doi.org/10.1007/s11214-015-0139-x.

[32] Hassler, M. D., Norbury, J. W., Reitz, G. (2017). Mars Science Laboratory Radiation Assessment Detector (MSL/RAD) Modeling Workshop Proceedings. *Life Sciences in Space Research*, 14. https://doi.org/10.1016/j.lssr.2017.06.004.

[33] Chian, A., *et al.* (2024). Terrestrial and Martian space weather: A complex systems approach. *Journal of Atmospheric and Solar-Terrestrial Physics*, 259, 106253. https://doi.org/10.1016/j.jastp.2024.106253.

Chapter 5

Health Risks of Ionizing Radiation

5.1 Introduction

Space radiation poses significant health risks to astronauts, given the unique and harsh environment of outer space. This type of radiation consists mainly of high-energy particles, such as protons, helium nuclei, and heavier ions, which can penetrate biological tissues and cause various types of damage. Understanding the nature and effects of space radiation is critical for ensuring the safety and well-being of astronauts on long-duration missions. This chapter provides an overview of radiobiological damage, including biomarkers that indicate such damage, and discusses the deterministic and stochastic effects of space radiation on human health.

Radiobiological damage refers to the harm caused to biological tissues by radiation. This damage can manifest at the molecular, cellular, and tissue levels, potentially leading to significant health issues. The extent and type of damage depend on factors such as the type of radiation, its energy, and the duration of exposure. Space radiation includes galactic cosmic rays (GCRs), solar particle events (SPEs), and radiation trapped within the Van Allen belts, each presenting different challenges to biological systems.

5.2 The Cascade of Radiation-Induced Damage

Radiobiology examines the complex interplay between radiation and living systems, highlighting how damage begins immediately upon interaction and evolves through several scientific domains. This sequence can be

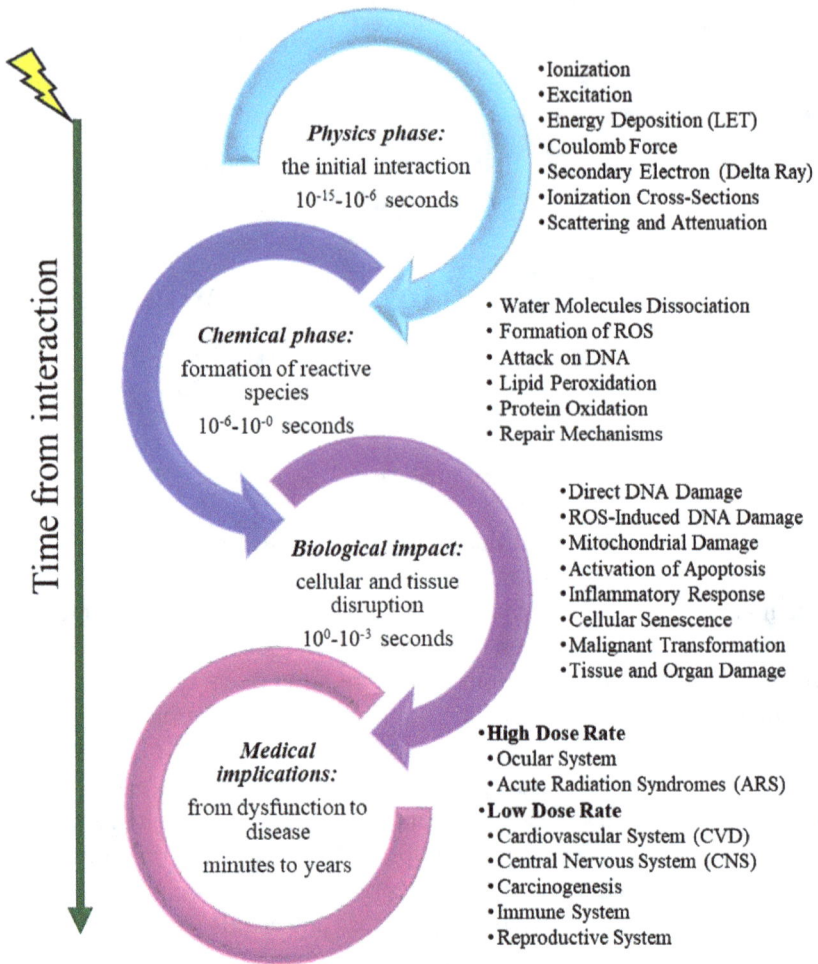

Fig. 5.1.   The ionizing radiation effects cascade.

Source: Generated with licensed MS Copilot tool by the authors.

categorized into four key phases, each occurring on different time scales: physics (femtoseconds), chemistry (milliseconds), biology (seconds to minutes), and medicine (minutes to years and next generation) (see Fig. 5.1).

5.2.1   *Physics phase: the initial interaction*

When radiation interacts with biological matter, the damage starts immediately through physical processes governed by fundamental principles.

Radiation, such as charged particles or electromagnetic waves, transfers energy to the atoms and molecules in its path. This interaction initiates several key phenomena (see Table 5.1), the most prominent being ionization, which refers to radiation particles colliding with atoms in biological matter and knocking electrons out of their orbitals. This process leaves behind positively charged ions and free electrons, forming "ion pairs." Also, *excitation* is possible since, in some cases, radiation excites the atom instead of ejecting it, raising its electrons to a higher energy state. While less disruptive than ionization, excitation alters the energy balance within molecules.

Generally, the basic effects derive from the energy deposition of the particle as it moves through matter in the medium. This energy transfer is usually quantified as *linear energy transfer* (LET) and depends on the type and energy of the radiation. High-LET radiation causes densely packed ionization tracks, leading to localized damage.

Table 5.1. Key phenomena and laws in the physics phase of ionizing radiation interaction with biological matter.

Phenomenon/Law	Description	Key Role in Radiation Damage
Ionization	Ejection of electrons from atoms due to radiation interaction.	Creates ion pairs, initiating molecular damage.
Excitation	Elevation of electrons to higher energy states without ejection.	Alters molecular energy states, priming chemical changes.
Energy Deposition (LET)	Energy is transferred by radiation per unit distance traveled in matter.	Determines the density and localization of ionization tracks.
Coulomb Force	The electrostatic force between charged particles.	Drives ionization by transferring energy from radiation to electrons.
Secondary Electron Production (Delta Ray)	High-energy ejected electrons ionize other atoms.	Amplifies initial damage, extending the affected area.
Ionization Cross-Sections	Probability of interaction between radiation and specific atoms/molecules.	Defines interaction likelihood and damage patterns.
Scattering and Attenuation	Alteration of radiation trajectory and energy loss during interaction with matter.	Influences penetration depth and distribution of radiation damage.

Also, a basic physical phenomenon can be described in terms of the Coulomb force, the force governs the interaction between charged radiation particles and electrons in biological matter. This fundamental force drives the attraction or repulsion between charges and influences ionization patterns.

Secondary electron production is also a typical phenomenon. It is similar to the secondary particle production described in the previous chapters for the interaction of cosmic rays with shielding materials or planetary regolith. In that case, ejected electrons, usually called "delta rays," carry sufficient energy to ionize nearby atoms. These secondary interactions amplify the damage and increase the range of affected regions.

Further, as radiation travels through matter, it may scatter off atoms or lose energy, thereby reducing its intensity and altering its trajectory. Energy conservation laws govern this behavior and are crucial in determining the penetration depth of radiation.

These processes occur on femtosecond timescales, creating a dynamic environment in which energy transfer mechanisms shape the initial damage sites in biological systems.

5.2.2   *Chemical phase: formation of reactive species*

Following physical interactions, the ionized molecules and radicals engage in chemical reactions. These processes, governed by chemical kinetics, form reactive oxygen species (ROS) and other highly reactive intermediates. Within milliseconds, these species can initiate damage to cellular components, including DNA, proteins, and lipids.

An exhaustive treatment of the chemical reactions in cells triggered by radiation goes beyond the scope of this book. Still, in general, they are bound to the ionization of water molecules in the human body and, consequently, the generation of free electrons. The ejected free electron can participate in other reactions, for example, in combination with ROS, generating attacks on different biomolecules, including cells' DNA, lipids, and proteins. Some peculiar chemical reactions, which form the basis of repair mechanisms, act as antioxidants and attempt to neutralize the generated ROS. Table 5.2 briefly describes the steps that occur at the chemical level after the interaction of ionizing radiation (IR) with biological matter.

In summary, water, the most abundant molecule in biological systems, produces ROS and is the primary target of IR. Reactive species such as OH, H_2O_2, and O_2 mediate the chemical damage. Further, DNA, lipids, and proteins are primary targets, with damage manifesting as strand

Table 5.2.   The sequence of reactions and their consequences for cellular damage, providing a concise overview of the chemical phase of radiation-induced effects.

Step	Reaction	Key Products	Impact on Cells
1. Ionization of Water	$H_2O \xrightarrow{\text{(radiation)}} H_2O^+ + e^-$	H_2O^+, e^-	Creates ionized water and free electrons, initiating damage.
2. Electron Hydration	$e^- \xrightarrow{\text{(hydration)}} e^-_{aq}$	e^-_{aq}	Alters molecular energy states, priming chemical changes.
3. Water Molecules Dissociation	$H_2O \xrightarrow{\text{(radiation)}} H^+ + OH\cdot$	H^+, H_2O_2, $OH\cdot$	Generates hydroxyl radicals, a major damaging agent.
4. Formation of ROS	$O_2 + e^-_{aq} \to O_2^-$ $O_2^- + H_2O_2 \to OH^- + OH + O_2$	O_2^-, $OH\cdot$	Produces reactive oxygen species that amplify damage.
5. Attack on DNA	$DNA + OH\cdot \to DNA\ damage$	Strand breaks, base modifications	Leads to mutations or cell death.
6. Lipid Peroxidation	$LH + OH\cdot \to L\cdot + H_2O$ $L + O_2 \to LOO\cdot$	Lipid radicals, lipid peroxides	Damages cell membranes, causing leakage and dysfunction.
7. Protein Oxidation	$Protein + OH\cdot \to Oxidized\ Protein$	Oxidized amino acids	Alters protein structure and function, disrupting processes.
8. Repair Mechanisms	$2O_2^- + 2H^+ \xrightarrow{\text{(SOD)}} H_2O_2 + O_2$ $2H_2O_2 \xrightarrow{\text{(catalase)}} 2H_2O + O_2$	Water and oxygen	Mitigates damage via antioxidants and repair enzymes (e.g., SOD).

breaks, mutations, lipid peroxidation, and protein dysfunction. Cellular repair systems can mitigate some damage, but excessive ROS overwhelms these mechanisms, leading to cell death or dysfunction.

5.2.3 *Biological impact: cellular and tissue disruption*

Chemical damage translates into biological consequences. Depending on the extent of the initial damage, cells may experience mutations, chromosomal aberrations, or apoptosis. The biological phase unfolds over minutes to days as the cell's repair mechanisms respond to the damage. When damage overwhelms these systems, it leads to dysfunction at the cellular and tissue levels. Table 5.3 summarizes the key events and effects in this biological phase of IR action, while Table 5.4 summarizes the relevant biomolecules affected.

In general, at the biological level, IR induces DNA damage and triggers corresponding repair mechanisms such as non-homologous end joining (NHEJ) or homologous recombination (HR), which, in the case of failure or inaccurate repair, can result in mutations, chromosomal aberrations, or apoptosis [1, 2]. Another path of damage arises when a high level of ROS overwhelms the cell; antioxidant defenses, such as glutathione and catalase, tip the balance toward oxidative stress. In that case, this phenomenon could trigger programmed cell death pathways such as apoptosis, autophagy, or necrosis in extreme cases. Also, in the same type of cellular tissue, chronic oxidative stress and DNA damage lead to long-term effects such as cancer, cardiovascular disease (CVD), and neurodegenerative disorders or damage to the central nervous system (CNS) can generate fibrosis due to excessive tissue remodeling, impairing organ function.

In the biological phase of radiation-induced damage, the chemical changes caused by ROS and other byproducts interact with vital biological molecules, such as DNA, proteins, and lipids. This phase leads to structural and functional alterations in cells, tissues, and organs. The outcomes of this interaction include mutations, cell death, inflammation, and impaired organ function, which can manifest as acute or chronic health effects.

5.2.4 *Medical implications: from dysfunction to disease*

The cumulative effects of cellular and tissue disruption manifest as medical conditions at the organ and systemic levels. These can range from

Table 5.3. Key events in the biological phase of IR action and their relevance for health in the short and long terms.

Step	Description	Biological Impact	Relevance to Health
1. Direct DNA Damage	High-energy radiation directly breaks DNA strands or modifies nucleotide bases.	Double-strand breaks (DSBs), single-strand breaks (SSBs), and base damage.	DSBs are the most lethal form of DNA damage, potentially causing cell death or mutations if not repaired properly.
2. ROS-Induced DNA Damage	ROS, such as OH•, attack DNA, causing oxidative base modifications and strand breaks.	8-oxoG formation, cross-links, and fragmentation of DNA strands.	Oxidative lesions can lead to mutations during replication, promoting cancer and other genetic disorders.
3. Lipid Peroxidation	ROS attacks membrane lipids, initiating a chain reaction of lipid oxidation.	Membrane damage, loss of integrity, increased permeability, and leakage of cellular contents.	Disruption of cellular membranes can lead to cell death and inflammation, affecting tissue and organ function.
4. Protein Oxidation	ROS oxidizes amino acids, alters protein structure, and forms carbonyl groups.	Enzyme inactivation, misfolding of proteins, and degradation.	Loss of protein function impairs cellular metabolism and stress responses, contributing to diseases such as Alzheimer's or Parkinson's.
5. Mitochondrial Damage	Radiation-induced ROS impairs mitochondrial DNA (mtDNA) and disrupts the electron transport chain.	Reduced ATP production, increased ROS generation, and mtDNA mutations.	Mitochondrial dysfunction contributes to apoptosis, energy crisis, and aging-related diseases.
6. Activation of Apoptosis	DNA damage or ROS-triggered signaling activates programmed cell death pathways.	Caspase activation, chromatin condensation, and cell shrinkage.	Apoptosis prevents the survival of severely damaged cells but may contribute to tissue dysfunction if excessive.

(Continued)

Table 5.3. (*Continued*)

Step	Description	Biological Impact	Relevance to Health
7. Inflammatory Response	Damaged cells release inflammatory signals, recruiting immune cells to the affected area.	Cytokine release, swelling, and further ROS production by immune cells.	Chronic inflammation can exacerbate tissue damage and is linked to diseases such as cancer, fibrosis, and cardiovascular disorders.
8. Cellular Senescence	Irreversible cell-cycle arrest occurs due to persistent DNA damage or oxidative stress.	Cellular aging, secretion of pro-inflammatory factors (SASP).	Senescent cells promote chronic inflammation and tissue degeneration, contributing to age-related and degenerative diseases.
9. Malignant Transformation	Accumulation of mutations leads to deregulation of cell growth and survival pathways.	Uncontrolled proliferation, angiogenesis, and evasion of apoptosis.	Malignant transformation leads to the formation of cancerous tumors, which are a long-term risk of radiation exposure.
10. Tissue and Organ Damage	Cell death, inflammation, and fibrosis combined effects impair tissue and organ function.	Loss of structural integrity, fibrosis, and organ failure.	Damaged tissues and organs can result in long-term health effects, including radiation syndromes, chronic diseases, and reduced functional capacity.

Table 5.4.   Summary of key biomolecules affected by radiation.

Biomolecule	Type of Damage	Cellular Consequences	Long-Term Effects
DNA	Strand breaks, base modifications, cross-links	Mutations, chromosomal aberrations, cell death	Cancer, genetic instability
Lipids	Peroxidation, loss of unsaturation	Membrane dysfunction increased permeability	Chronic inflammation, organ dysfunction
Proteins	Oxidation, fragmentation, misfolding	Enzyme inactivation, loss of structural integrity	Impaired metabolism, neurodegenerative diseases
Mitochondria	ROS production, mtDNA damage	Energy crisis, activation of apoptosis	Aging, metabolic disorders
Cytoskeleton	Cross-linking, fragmentation	Loss of cell shape, impaired intracellular transport	Tissue weakness, impaired repair mechanisms

acute radiation syndromes (ARSs) to long-term consequences, such as cancer, fibrosis, or organ failure. Medicine addresses these challenges by diagnosing, treating, and mitigating the effects of radiation-induced damage, focusing on restoring health and functionality [3].

During space missions and in environments with high radiation exposure, the human body is susceptible to various adverse effects spanning different organ systems. These effects vary in severity and can manifest as acute or long-term health consequences.

In the ocular system, astronauts often report experiencing visual phenomena known as eye flashes, bright streaks or flashes of light perceived during missions. These are thought to result from high-energy particles, such as cosmic rays, interacting with retinal photoreceptors or stimulating the optic nerve. Over prolonged exposure, radiation can also lead to the development of cataracts. This condition arises from structural changes in the lens, where radiation-induced damage to epithelial cells triggers protein aggregation and lens cloudiness, ultimately impairing vision. Cataracts are a delayed but significant risk for astronauts and radiation workers.

Radiation exposure also affects CVD by causing damage to blood vessels. Chronic exposure can induce inflammation, promote atherosclerosis, and lead to fibrosis, thereby increasing the risk of heart disease and stroke. These effects concern individuals exposed to high doses over extended periods, such as astronauts on deep-space missions or patients undergoing radiotherapy.

The CNS is another critical area of concern. Exposure to low or moderate doses of radiation over time can impair cognitive functions, memory, and motor coordination. This damage is primarily caused by oxidative stress and inflammation, which affect neurons and supporting glial cells. At extremely high doses, radiation can lead to acute CNS syndrome, characterized by severe and rapid tissue damage, swelling, and bleeding. This condition is fatal within hours or days if the exposure exceeds tolerable levels. One of the most severe long-term consequences of radiation exposure is carcinogenesis. The DNA damage and genomic instability induced by radiation increase the likelihood of mutations that can accumulate over time, eventually leading to cancer [4]. Common cancers associated with radiation exposure include leukemia, as well as solid tumors in organs such as the thyroid, breast, and skin.

For individuals exposed to high doses of radiation, a spectrum of ARSs can occur. A hematopoietic syndrome is marked by suppressed bone marrow activity, resulting in reduced blood cell production, anemia, heightened susceptibility to infections, and impaired clotting. Gastrointestinal syndrome, on the other hand, involves severe damage to the intestinal lining, causing nausea, vomiting, diarrhea, fluid loss, and a high risk of sepsis. In extreme cases, neurovascular syndrome may develop, characterized by the rapid destruction of brain tissue and vasculature, leading to a loss of coordination, seizures, and death within hours to days.

Other significant health risks include reproductive damage, where radiation affects germ cells, potentially causing infertility or genetic defects in offspring. Additionally, immune dysfunction can occur, as radiation impairs the production and functionality of immune cells, increasing susceptibility to infections.

Table 5.5 highlights the clinical effects of IR, dividing them into the human body system and giving relevance to the rate of IR exposition, or, in other words, whether the amount of radiation is absorbed in a short amount of exposure time (high dose rate in minutes to hours) or a long one (months or years). In the first case, the effects or syndromes are usually "deterministic," with the other usually referred to as "stochastic." In the medical phase of radiation damage, the biological consequences manifest as clinical symptoms, syndromes, or long-term health risks. These effects vary depending on the radiation dose, type, exposure duration, and individual susceptibility. The medical phase focuses on understanding, diagnosing, and managing these health risks to minimize their impact on

Table 5.5. Low-dose-rate effects on human health.

System Affected	Effect	Mechanism	Clinical Relevance
Cardiovascular System	*Cardiovascular disease* (**CVD**): Increased risk of heart disease and stroke.	Endothelial damage, inflammation, and fibrosis in blood vessels due to chronic radiation exposure.	Long-term risk for astronauts and cancer patients treated with radiotherapy; a significant concern for high-dose exposure scenarios.
Central Nervous System (CNS)	*Neurocognitive decline:* Memory, cognition, and motor skill impairments.	Oxidative stress and inflammation damage neurons and glial cells, impairing neurogenesis.	Long-term exposure to space radiation (e.g., galactic cosmic rays) is linked to cognitive deficits and neurodegeneration.
Carcinogenesis	*Cancer development:* Increased incidence of leukemia, solid tumors, and other malignancies.	DNA mutations, genomic instability, and failure of repair mechanisms.	Long-term effects depend on dose, type, and exposure duration. Common cancers include lung, breast, thyroid, and skin cancers.
Immune System	*Increased susceptibility to infections*	Chronic low-dose radiation leads to immune dysregulation.	Stochastic risk, especially in long-term exposure scenarios (e.g., space missions).
	Immune-related diseases	Long-term effects, such as autoimmunity or chronic inflammation.	Probabilistic effect linked to cumulative low-dose exposure.
Reproductive System	*Congenital disabilities in offspring*	Radiation-induced mutations in germ cells (sperm/ova).	Stochastic risk with no threshold. Monitoring required for space missions.
	Genetic mutations	Random DNA damage in reproductive cells.	Long-term risk for subsequent generations.

human health. In general, each system can be affected either deterministically or stochastically, with the difference primarily lying in the type of exposure. Deterministic effects refer to effects that always occur after a certain threshold of exposure to IR or absorbed dose is reached. In contrast, stochastic ones involve random, probabilistic damage without a requisite minimum dose. Of course, the probability of stochastic effects increases with the dosage of absorbed radiation.

Table 5.5 summarizes the possible clinical effects induced by high levels of ionizing radiation exposure over a long time due to the presence of low-dose-rate radioactive sources. These effects are usually referred to as stochastic. Stochastic effects involve random, probabilistic damage, with no minimum dose required. The probability of such effects increases with dosage, but their severity is not dose-dependent.

Table 5.6 summarizes the possible clinical effects induced by high levels of IR exposure in a short time due to high-dose-rate radioactive sources. These effects are usually referred to as deterministic.

5.3   Tracing Radiation Damages: Biomarkers

Biomarkers are measurable indicators of biological processes, states, or conditions, and they are crucial for assessing the extent of radiobiological damage. Biomarkers can be found in blood, urine, tissues, and even hair, providing vital information about the biological impact of radiation exposure. In space activities where the support of professional medical healthcare could be reduced or limited in time, it will also become paramount to select and improve knowledge of potential indicators of the levels of exposure in real time using pieces of information that can also allow modifying the profile or time duration of a space mission for the entire crew as well as individual astronauts/space workers. Figure 5.2 shows an overview of biomarkers associated with ionizing radiation mechanisms, categorized by their cellular location. In this section, we review the most common biomarkers used so far and the perspective of their utilization in deep-space exploration [5].

5.3.1   *DNA modifications as biomarkers of ionizing radiation effects*

DNA is the fundamental blueprint of life, encoding the genetic instructions that guide the development, functioning, growth, and reproduction of all living organisms, including humans. It is a double-stranded molecule

Table 5.6.   High-dose-rate effects on human health.

System Affected	Effect	Mechanism	Clinical Relevance
Ocular System	*Eye flashes:* Perception of light flashes due to radiation interaction with the retina.	Charged particles (e.g., cosmic rays) stimulate photoreceptors or optic nerve pathways.	Observed in astronauts during space missions; not harmful but indicative of radiation exposure.
	Cataracts: Clouding of the eye lens. Mostly deterministic effects.	Radiation damages lens epithelial cells, causing protein aggregation and opacity.	Delayed effect: common in astronauts and radiation workers, leading to vision impairment.
Acute Radiation Syndromes (ARS)	*Acute CNS syndrome:* At very high doses, seizures, loss of consciousness, and death.	Severe damage to neurons and blood–brain barrier; rapid swelling and hemorrhage.	Observed in extreme exposure scenarios (>50 Gy).
	Hematopoietic syndrome: Bone marrow suppression, anemia, and infections.	Radiation kills rapidly dividing bone marrow cells, reducing blood cell production.	Observed in doses >1 Gy; can be lethal without treatment (bone marrow transplant, antibiotic).
	Gastrointestinal syndrome: Nausea, vomiting, diarrhea, and dehydration.	Losing intestinal lining cells leads to ulceration, fluid loss, and sepsis.	Observed in doses >6 Gy; severe cases require intensive medical care.
	Vascular syndrome: Loss of coordination, seizures, and shock.	Vascular tissues are rapidly destroyed at high doses (>20 Gy).	Fatal within hours to days; no effective treatment.

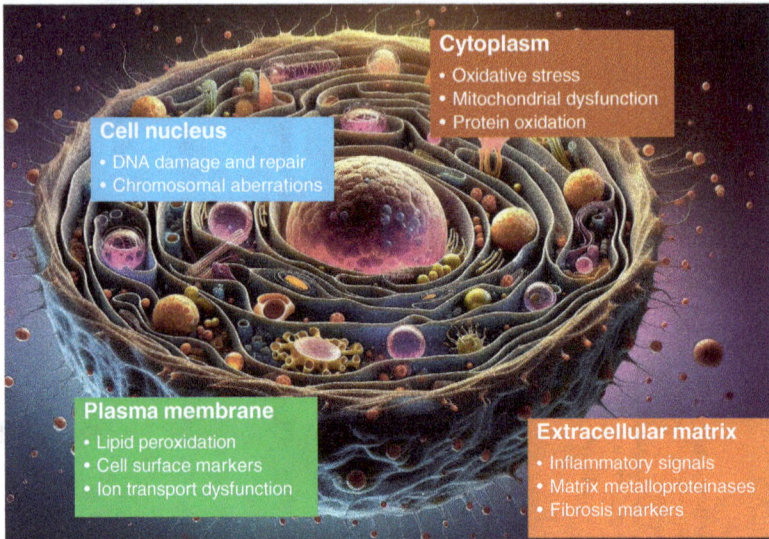

Fig. 5.2. An overview of biomarkers associated with ionizing radiation mechanisms, categorized by their cellular location. These biomarkers indicate radiation-induced damage and responses in the nucleus (e.g., DNA damage markers), cytoplasm (e.g., oxidative stress indicators), plasma membrane (e.g., lipid peroxidation markers), and extracellular matrix (e.g., inflammatory and fibrosis markers).

Source: Generated with licensed MS Copilot tool by the authors.

composed of four nucleotide bases, adenine (A), thymine (T), cytosine (C), and guanine (G), arranged in specific sequences that form the genetic code. These sequences determine the synthesis of proteins essential to cellular structure and function. The strands of DNA are held together by hydrogen bonds between complementary base pairs, with A pairing with T and C pairing with G, forming the iconic double-helix structure.

In humans, DNA is organized into highly structured entities known as chromosomes, which reside in the nucleus of each cell. Humans typically have 46 chromosomes arranged in 23 pairs. Of these, 22 pairs are autosomes, and one pair consists of sex chromosomes, which determine biological sex (XX for females and XY for males). Each chromosome contains a single, long molecule of DNA intricately packaged with histone proteins to form chromatin. This organization ensures the efficient storage of genetic material and regulates access to DNA during processes such as replication and transcription.

Chromosomes also feature specialized regions crucial for their function. Telomeres at the ends of chromosomes protect DNA from degradation and prevent chromosomal fusions. The centromere, a constricted region, plays a key role during cell division by anchoring the spindle fibers that segregate chromosomes into daughter cells.

Given its central role in cellular function and heredity, DNA is also the primary target of damage induced by environmental stressors, including IR. Such damage can manifest in various forms, including single-strand breaks (SSBs), double-strand breaks (DSBs), base modifications, and chromosomal aberrations. Studying these molecular and structural changes provides insight into the potential biomarkers of IR exposure, aiding in our understanding of radiation-induced effects on human health.

5.3.2 *DNA breaks*

One of the primary indicators of radiobiological damage is DNA breaks. These breaks can be SSBs or DSBs, with the latter being more challenging to repair and more likely to lead to mutations or cell death. DNA breaks directly result from the high-energy particles interacting with the DNA molecules, causing physical disruptions in the DNA structure.

SSBs break one of the two DNA strands, which the cell's natural repair mechanisms can typically repair. However, if left unrepaired or mis-repaired, SSBs can lead to significant cellular dysfunction. On the other hand, DSBs are more severe, involving the breaking of both DNA strands. These breaks are more challenging to repair and are often associated with higher risks of mutations and carcinogenesis. The repair of DSBs involves complex mechanisms, such as homologous recombination and non-homologous end joining, which, if erroneous, can result in mutations and chromosomal aberrations.

Studies have shown that high-LET radiation encountered in space is more potent at causing DSBs than low-LET radiation on Earth, such as X-rays. The capability of high-LET radiation in causing DSBs underscores the importance of developing effective protective measures and therapeutic strategies to mitigate these effects during space missions.

Advances in molecular biology have provided techniques for detecting and quantifying DNA breaks across different tissues and cell types. These methods are pivotal for understanding radiation-induced damage and tailoring countermeasures. Key techniques include the following:

- *Comet assay (single-cell gel electrophoresis)*: Sensitive for detecting SSBs and DSBs at the single-cell level.

- *γ-H2AX foci assay*: DSBs are identified through immunofluorescence tagging phosphorylated H2AX histones at DNA break sites.
- *Neutral and alkaline gel electrophoresis*: Differentiates between SSBs and DSBs by analyzing DNA fragmentation.
- *Quantitative PCR and next-generation sequencing (NGS)*: Quantifies DNA break frequency and repair efficiency.

Each technique could be more effective in different cell tissues, as highlighted in Table 5.7.

Table 5.7. Summary of DNA damage response across various tissue and cell types.

Tissue/Cell Type	Characteristics	Techniques for Analysis	Key Observations
Lymphocytes	High sensitivity to radiation; commonly studied *in vitro*	γ-H2AX foci assay, comet assay	Rapid detection of DSBs and SSBs.
Neurons	Limited repair capacity due to non-dividing nature	γ-H2AX foci assay, immunohistochemistry	Persistent damage linked to neurodegeneration.
Epithelial Cells	Rapidly dividing, efficient repair mechanisms	Neutral gel electrophoresis, comet assay	High repair activity but prone to errors in NHEJ.
Fibroblasts	Proficient in repair; used for wound-healing studies	Quantitative PCR, next-generation sequencing (NGS)	Efficient DSB repair but affected by oxidative stress.
Cancer Cells	Often repair-deficient due to mutations	γ-H2AX foci assay, alkaline gel electrophoresis	Accumulation of unrepaired DSBs enhances therapeutic targeting.
Stem Cells	High repair fidelity; critical for regenerative capacity	Single-cell sequencing, comet assay	Radiation exposure affects differentiation potential.

This table summarizes the DNA damage response across various tissue and cell types, highlighting their unique characteristics, preferred techniques for DNA break analysis, and key observations regarding their sensitivity to IR. This comparison underscores the variability in DNA repair capacity and the impact of radiation on various biological systems.

5.3.3 *Base modifications and chromosomal aberrations*

Base modifications refer to damage to the base elements of DNA. Base modifications involve chemical changes to the DNA bases (e.g., oxidation, alkylation, and deamination) without breaking the DNA strand. Alterations occur in the base-pairing properties or the base's ability to be recognized during transcription and replication. The leading cause is ROS or radiation that modifies the chemical structure of the base. The typical repair mechanisms are base excision repair (BER) and nucleotide excision repair (NER), which remove and replace damaged bases. If the damage is not repaired, it can result in single point mutations or base mispairing during replication, potentially leading to genetic mutations.

Chromosomal aberrations are structural changes in chromosomes resulting from radiation exposure. These changes can include deletions, duplications, inversions, and translocations of chromosome segments. Chromosomal aberrations are significant because they can lead to genomic instability, cancer, and other genetic disorders [6, 7].

Deletions involve the loss of a chromosome segment, while duplications involve the repetition of a segment. Inversions refer to the flipping of a chromosome segment within the same chromosome, and translocations involve the exchange of segments between non-homologous chromosomes. These structural changes can disrupt gene function and regulation, leading to various diseases, including cancer. It is possible to distinguish aberrations into two general categories: *simple* and *complex*. Simple aberrations involve a single chromosome, while complex aberrations involve multiple ones.

Tables 5.8 and 5.9 describe the main types of chromosomal aberrations, distinguishing simple from complex, while Table 5.8 compares some features of base modifications, SSBs, DSBs, and chromosomal aberrations.

Table 5.8.   Overview of simple chromosomal aberrations induced by IR.

Type of Aberration	Description	Mechanism of Formation	Potential Consequences
Deletions	Loss of a chromosomal segment	Misrepair of double-strand breaks (DSBs), failure in homologous recombination	Loss of essential genetic material, potentially leading to gene inactivation
Duplications	Duplication of a chromosomal segment	Unequal cross-over during repair or replication errors	Overexpression of duplicated genes, contributing to genomic instability
Inversions	Reversal of a chromosomal segment within the same chromosome	Incorrect repair of DSBs, misalignment during homologous recombination	Disruption of gene function, formation of fusion genes
Chromatid Aberrations	Damage restricted to one chromatid, such as breaks or exchanges	Radiation exposure during the S or G2 phase of the cell cycle	Structural anomalies upon subsequent division

Table 5.8 presents an overview of simple chromosomal aberrations induced by IR, including deletions, duplications, inversions, and chromatid aberrations. These changes are localized to a single chromosome or chromatid and primarily result from errors in the repair of DNA DSBs or replication errors.

Table 5.9 presents an overview of IR-induced complex chromosomal aberrations, including translocations, aneuploidy, ring chromosomes, dicentric chromosomes, isochromosomes, fragmentation, and chromosome bridges. These aberrations involve multiple chromosomes or significant structural rearrangements, often resulting in severe genomic instability and cellular dysfunction.

A comparative overview of DNA and chromosomal damage types is given in Table 5.10, organized by increasing complexity of structural disruption. The table highlights key features, causes, repair mechanisms, severity, and detection techniques for base modifications, SSBs, DSBs,

Table 5.9. Overview of complex chromosomal aberrations induced by IR.

Type of Aberration	Description	Mechanism of Formation	Potential Consequences
Translocations	Exchange of segments between two non-homologous chromosomes	Non-homologous end joining (NHEJ) of DSBs	Altered gene regulation, activation of oncogenes, or inactivation of tumor suppressors
Aneuploidy	Abnormal number of chromosomes	Errors during mitotic spindle assembly or chromosome segregation	Genetic imbalance, often lethal or associated with cancer
Ring Chromosomes	Formation of a circular chromosome due to loss of telomeric regions	Misrepair of DSBs, a fusion of broken ends	Genetic instability, often associated with cell death or malignancy
Dicentric Chromosomes	Chromosome with two centromeres resulting from the fusion of two broken chromosome ends	Misrepair of DSBs, failure of proper chromosome segregation	Chromosomal instability, mitotic arrest, or cell death
Isochromosomes	Chromosomes with identical arms, resulting in loss of genetic material from one arm and duplication of the other	Misrepair during cell division	Gene dosage imbalance, associated with developmental disorders and cancer
Fragmentation	Chromosome shattered into multiple pieces	Extensive IR damage leading to multiple DSBs	Loss of genetic information is often lethal for the cell
Chromosome Bridges	Formation of a bridge between segregating chromosomes during anaphase	Fusion of chromosome ends, typically telomere dysfunction	Breakage during mitosis, leading to further chromosomal instability

Table 5.10.   Comparative overview of DNA and chromosomal damage types.

Feature	Single-Strand Breaks (SSBs)	Base Modifications	Double-Strand Breaks (DSBs)	Chromosomal Aberrations
Nature of Damage	Break in one strand of DNA	Chemical alteration of a DNA base	Break in both DNA strands	Structural changes in chromosomes
Impact on DNA Structure	Backbone discontinuity	No break in the backbone	Complete breakage of the double helix	Large-scale rearrangements or losses
Primary Cause	IR, oxidative stress	Oxidative or chemical agents	High-energy radiation or replication stress	DSB misrepair, replication stress, or radiation
Repair Mechanism	SSBR or BER	BER or NER	Homologous recombination (HR) or non-homologous end joining (NHEJ)	No direct repair; results from DSB misrepair
Potential Consequences	Leads to DSBs if unrepaired	Point mutations or mispairing	Mutations, cell death, carcinogenesis	Loss of genetic material, aneuploidy, or tumorigenesis
Severity of Damage	Moderate	Low to moderate	High	Very high
Detection Techniques	Comet assay, alkaline gel electrophoresis	Mass spectrometry, BER-specific assays	γ-H2AX foci assay, neutral gel electrophoresis	Karyotyping, spectral karyotyping (SKY), fluorescence *in situ* hybridization (FISH)

and chromosomal aberrations. This structured comparison emphasizes the escalating impact of damage on genomic integrity and cellular health.

5.3.4 *MiRNA*

MicroRNAs (miRNAs) are small, non-coding RNA molecules that play crucial roles in gene regulation. Radiation can alter the expression of miRNAs, which, in turn, can affect various cellular processes, including apoptosis, cell cycle regulation, and DNA repair mechanisms [8].

MiRNAs are 19–25 nucleotides long and function by binding to messenger RNAs (mRNAs), leading to their degradation or the inhibition of translation. This post-transcriptional regulation is vital for maintaining cellular homeostasis. Radiation-induced changes in miRNA expression can disrupt these regulatory networks, contributing to cellular dysfunction and disease.

Studies [8–10] have shown that specific miRNAs are upregulated or downregulated in response to radiation exposure. For example, miR-21, miR-34a, and miR-146a have been identified as radiation-responsive miRNAs involved in processes such as apoptosis and inflammation. These miRNAs can serve as biomarkers for radiation exposure and potential therapeutic targets for mitigating radiation-induced damage. Another critical aspect of miRNA as a biomarker of IR is its presence in solid tissues and various body fluids (e.g., serum, urine, and saliva).

The role of miRNAs in radiation response is an active area of research. Studies are ongoing to elucidate the molecular mechanisms underlying miRNA regulation by radiation and their impact on cellular and tissue function. Understanding these mechanisms will help develop miRNA-based radiation protection and therapy strategies for space missions.

5.3.5 *Micronuclei*

Micronuclei are small nuclei that form in cells when chromosome fragments or whole chromosomes are not incorporated into the daughter nuclei during cell division [11, 12]. The presence of micronuclei is a marker of genomic instability and has been associated with radiation exposure.

Micronuclei can result from chromosomal breakage or malfunction of the mitotic spindle apparatus during cell division. They indicate underlying genetic damage and are used as biomarkers for assessing the genotoxic effects of radiation.

The micronucleus assay is a widely used technique for detecting micronuclei and evaluating the genotoxic potential of radiation. This assay involves staining cells and examining them under a microscope to identify and count micronuclei. An increased frequency of micronuclei in cells indicates higher levels of genomic instability and radiation-induced damage.

Research has shown that astronauts exposed to space radiation exhibit an increased frequency of micronuclei in their lymphocytes, highlighting the genotoxic impact of space radiation. Monitoring micronuclei in astronauts provides valuable information on their radiation exposure and helps assess the effectiveness of protective measures.

5.3.6 *Other biomarkers*

Other potential biomarkers of radiobiological damage include changes in protein expression levels, metabolic profiles, and specific molecular signatures in blood and tissues. Ongoing research continues to identify and validate new biomarkers that can help in the early detection and monitoring of radiation-induced damage.

Proteomics [13], the study of the entire set of proteins expressed by a cell or tissue, provides insights into the changes in protein expression in response to radiation. Radiation can induce the expression of stress response proteins, DNA repair proteins, and inflammatory proteins. These changes can serve as biomarkers for assessing the biological impact of radiation exposure. The use of proteins as biomarkers during space missions is under investigation. Biomarkers such as γ-*H2AX* and *SOD* are particularly promising due to their rapid response to radiation and potential for integration with miniaturized detection platforms. In contrast, others, such as *IL-6* and *TGF-β1*, can provide broader insights into systemic health effects, including inflammation and tissue remodeling, which are essential for long-term missions. In general, the use of these proteins in space will depend on advancements in portable diagnostic technologies, such as lab-on-a-chip systems and compact ELISA platforms, which will enable real-time or near-real-time biomonitoring [14]. Table 5.11 provides

Table 5.11. Protein biomarkers for IR damage, emphasizing their functional relevance and potential applicability for space missions.

Protein	Function/Role	Relevance to IR Damage	Use in Space Missions
γ-H2AX	Phosphorylated histone variant involved in DNA damage response	Marker of DNA double-strand breaks (DSBs); rapidly forms foci at DSB sites	Reliable for monitoring radiation-induced DNA damage in space; may require miniaturized detection systems
SOD (Superoxide Dismutase)	Antioxidant enzyme neutralizing superoxide radicals	Protects cells from radiation-induced oxidative stress	Practical for monitoring oxidative stress in astronauts; potential integration with portable biosensors
IL-6 (Interleukin-6)	Pro-inflammatory cytokine	Elevated levels associated with radiation-induced inflammation and tissue damage	Capable of monitoring astronauts' inflammation and immune system responses during space missions
TGF-β1	Transforming growth factors involved in fibrosis and repair	Elevated levels indicate radiation-induced tissue remodeling and fibrosis	Relevant for assessing long-term tissue damage and fibrosis risks in prolonged space missions
p53	Tumor suppressor protein regulating cell cycle and apoptosis	Activates in response to DNA damage; controls repair or apoptosis pathways	Helpful in assessing radiation-induced apoptosis; requires advanced analytical equipment
Rad51	Recombinase is critical for homologous recombination repair	Mediates DNA strand exchange during DSB repair; essential for accurate repair	Relevant for assessing radiation effects on DNA repair fidelity in microgravity conditions
Ku70/Ku80	Components of the non-homologous end joining (NHEJ) pathway	Bind to DNA ends at DSBs to initiate repair; critical for rapid, albeit error-prone repair	Help monitor DSB repair in space, especially under high-LET radiation exposure
ATM	Serine/threonine kinase regulating DNA repair and cell cycle	A central regulator of the DNA damage response phosphorylates downstream repair proteins	Provides insights into the efficiency of DNA repair pathways under space radiation conditions

Table 5.12.   Comparative analysis of biomarkers: miRNA, micronuclei, proteomics, and metabolomics.

Biomarker	Characteristics	Potential as Biomarker for IR	Drawbacks for Use in Space
miRNA	Small non-coding RNA molecules that regulate gene expression at the post-transcriptional level.	Altered expression profiles in response to radiation can indicate DNA damage, apoptosis, or cellular stress.	Limited knowledge of space radiation-specific miRNA profiles requires complex sequencing equipment.
Micronuclei	Small, extranuclear bodies containing chromosomal fragments or whole chromosomes are excluded during cell division.	Indicators of chromosomal instability and DNA damage; easily detectable using microscopy or flow cytometry.	Labor-intensive assays; sensitivity to low-dose radiation effects may be limited.
Proteomics	Study of the complete protein set expressed by an organism, including post-translational modifications.	Identifies radiation-specific protein expression patterns linked to DNA repair, oxidative stress, and apoptosis.	Requires sophisticated instruments, such as mass spectrometers, which may need to be more practical in space.
Metabolomics	Study of small molecules (metabolites) that reflect the biochemical activity within a cell or organism.	Tracks radiation-induced changes in energy metabolism, oxidative stress, and signaling pathways; highly sensitive.	Samples may degrade quickly without precise storage conditions; complex data analysis is required.

an overview of certain proteins that can be used in that sense, assessing potential use in deep-space missions. Metabolomics [15], the study of the complete metabolites within a biological sample, also provides valuable information on the metabolic changes induced by radiation. Radiation can alter metabolic pathways, leading to changes in energy production, lipid metabolism, and oxidative stress. These metabolic changes can serve as biomarkers for radiation exposure and potential therapeutic targets.

Identifying molecular signatures, such as specific RNA or DNA sequences, also holds promise as biomarkers for radiation exposure. Advances in genomics and high-throughput sequencing technologies enable the discovery of novel molecular signatures that can provide sensitive and specific indicators of radiobiological damage.

Table 5.12 compares the last categories approached in the chapter and highlights the potential drawbacks of their use during deep space missions.

By continuously advancing our understanding of biomarkers and developing new technologies for their detection, we can improve the monitoring and assessment of radiation exposure and its health effects, ultimately enhancing the safety and well-being of astronauts on long-duration space missions.

Table 5.12 provides a comparative analysis of biomarkers, miRNA, micronuclei, proteomics, and metabolomics, highlighting their characteristics, potential as indicators of IR effects, and challenges for implementation in space research. This evaluation emphasizes the strengths and limitations of each biomarker in monitoring radiation-induced biological changes during space missions.

References

[1] Goodhead, D. T. (1994). Initial events in the cellular effects of ionizing radiations: Clustered damage in DNA. *International Journal of Radiation Biology*, 65(1), 7–17.

[2] Ward, J. F. (1995). Radiation mutagenesis: The initial DNA lesions responsible. *Radiation Research*, 142(3), 362–368.

[3] Hall, E. J., and Giaccia, A. J. (2006). *Radiobiology for the Radiologist*. Lippincott Williams & Wilkins, Philadelphia, PA.

[4] Durante, M., and Cucinotta, F. A. (2008). Heavy ion carcinogenesis and human space exploration. *Nature Reviews Cancer*, 8(6), 465–472.

[5] Bentzen, S. M., Parliament, M., Deasy, J. O., Dicker, A., Curran, W. J., Williams, J. P., Rosenstein, B. S. (2010). Biomarkers and surrogate endpoints for normal-tissue effects of radiation therapy: The importance of dose–volume effects. *International Journal of Radiation Oncology Biology Physics*, 76(3 Suppl), S145–S150. https://doi.org/10.1016/j.ijrobp.2009.08.076.

[6] Natarajan, A. T., and Obe, G. (1983). Molecular mechanisms involved in the production of chromosomal aberrations: II. Rejoining of broken ends and their misrepair. *Mutation Research*, 110(2), 209–229.

[7] Thacker, J. (1992). The nature of mutations induced by ionizing radiation in cultured mammalian cells. *BioEssays*, 14(12), 831–838.

[8] Meng, J., Wang, Z. (2022). MicroRNAs as biomarkers for ionizing radiation injury. *Frontiers in Cell and Developmental Biology*, 10. https://doi.org/10.3389/fcell.2022.861451.

[9] Nicoloso, M. S., Spizzo, R., Shimizu, M., Rossi, S., and Calin, G. A. (2009). MicroRNAs – The micro steering wheel of tumour metastases. *Nature Reviews Cancer*, 9(4), 293–302.

[10] Iorio, M. V., and Croce, C. M. (2012). MicroRNA dysregulation in cancer: Diagnostics, monitoring and therapeutics. A comprehensive review. *EMBO Molecular Medicine*, 4(3), 143–159.

[11] Fenech, M. (2000). The in vitro micronucleus technique. *Mutation Research/Fundamental and Molecular Mechanisms of Mutagenesis*, 455(1-2), 81–95.

[12] Fenech, M. (2007). Cytokinesis-block micronucleus cytome assay. *Nature Protocols*, 2(5), 1084–1104.

[13] Hanash, S. M., Pitteri, S. J., and Faca, V. M. (2008). Mining the plasma proteome for cancer biomarkers. *Nature*, 452(7187), 571–579.

[14] Engvall, E., and Perlmann, P. (1971). Enzyme-linked immunosorbent assay (ELISA): Quantitative assay of immunoglobulin G. *Immunochemistry*, 8(9), 871–874. https://doi.org/10.1016/0019-2791(71)90454-x.

[15] Tyburski, J. B., Patterson, A. D., Krausz, K. W., Slavík, J., Fornace, A. J., and Gonzalez, F. J. (2010). Radiation metabolomics. 3. Biomarker discovery in the urine of γ-irradiated rats using global metabolomics profiling by UPLC-ESI-QTOFMS. *Radiation Research*, 173(2), 204–217.

Chapter 6

Radiation Health Risk Assessment for Humans in Space Missions

6.1 Introduction

Human exploration of space pushes the boundaries of science and technology, presenting unique challenges in ensuring astronaut safety and mission success. One of the most significant threats in space is exposure to ionizing radiation (IR), which can cause severe biological damage and long-term health consequences. In space, the absence of Earth's protective atmosphere and magnetic field leaves astronauts vulnerable to a spectrum of high-energy particles, including solar particle events (SPEs) and galactic cosmic rays (GCRs). Addressing this risk requires a multidisciplinary approach to assess, model, and mitigate the impact of radiation on the human body. This chapter explores this critical topic, offering a comprehensive framework for understanding and addressing radiation risks during space missions.

The chapter begins with Section 6.2, which introduces the foundations of radiation risk assessment through the lens of an ontology development approach. This novel perspective emphasizes systematic knowledge organization by integrating data from diverse fields, such as space physics, radiobiology, and computational modeling. By constructing a structured ontology, researchers and mission planners can better understand the complex interactions between space radiation and biological systems, thereby enabling more accurate risk predictions and targeted countermeasures.

Section 6.3 delves into the methods used to evaluate biological damage and its implications for risk assessment. The evaluation process begins with dose estimation and the concept of relative biological effectiveness (RBE), which quantifies how different types of radiation affect living tissues. Establishing dose limits is a cornerstone of astronaut safety. This section also explores NASA's protocols for protecting astronauts, focusing on the risk of fatal cancers caused by radiation exposure. Advanced modeling approaches, including the track structure model and cell survival probability functions, provide detailed insights into cellular damage and repair mechanisms. Furthermore, Monte Carlo (MC) computational simulations offer powerful tools for predicting radiation interactions at the microscopic level, enhancing the precision of risk assessment models.

Experimental studies validate theoretical models and refine our understanding of space radiation's biological effects. Section 6.4 explores the experimental aspect of space radiobiology, focusing on using biological models and controlled radiation exposure protocols. Subsections examine the design and application of biological models, protocols for simulating space-like radiation conditions on Earth, and the direct measurement of radiation effects in space. Additionally, the importance of accelerator infrastructure and specialized equipment for reproducing the high-energy particle environment of space is discussed, highlighting their indispensable role in advancing space radiobiology research.

This chapter integrates theoretical frameworks, advanced computational tools, and experimental studies to equip researchers, engineers, and policymakers with the knowledge and methodologies needed to protect astronauts and ensure the success of future missions.

6.2 Foundations of Risk Assessment for Ionizing Radiation in Space: An Ontology Development Approach

Assessing IR risks in space is foundational to ensuring astronaut safety and mission success, particularly for long-duration space exploration beyond Earth's magnetosphere. In that sense, to understand and address the complexities of space radiation risks, researchers can employ the more general concepts of ontologies, dimensions, and taxonomies as defined in the following:

- *Ontology*: A structured framework representing knowledge, ontologies define entities (e.g., concepts and objects) and their relationships.

They ensure consistency and interoperability in research, dataset management, and risk modeling.

- *Dimensions*: Thematic areas or aspects of a phenomenon, dimensions organize the core categories of investigation (e.g., study types, health effects, and environmental factors).
- *Taxonomies*: Hierarchical or categorical classifications within dimensions, taxonomies group related entities into logical clusters (e.g., types of radiation or health outcomes).

This structured approach supports cross-comparisons of data from diverse sources, enhances reliability in modeling and predictions, and creates a robust framework for long-term investigations.

An example of such methodology [1] is an ontology for space radiobiology, based on four dimensions ("Study Approach," "Medicine," "Radiation," and "Radiobiology Key Performance Indicators" (KPIs)). Inside these dimensions, 11 taxonomies were identified, offering a systematic framework for organizing knowledge and facilitating interdisciplinary research. The taxonomies are reported in the following paragraphs and depicted in Fig. 6.1.

- *Study approaches*: Risk assessment in space radiation relies on three primary models:
  - *Computational models*: Advanced algorithms simulate radiation interactions and biological outcomes, enabling predictions of exposure effects over time and in varying conditions.
  - *Mechanistic models*: Focus on the underlying biological processes, such as DNA damage and repair pathways, offering insights into how radiation affects cellular and molecular systems.
  - *Phenomenological models*: Utilize empirical data to describe observed radiation effects, often in dose–response relationships.
  - Research also integrates data from space missions and Earth-based studies, providing a diverse foundation for understanding the risks. Space missions supply real-world exposure data, while Earth-based analogs, such as ground-based particle accelerators, enable controlled experimentation.
- *Medicine*: The medical perspective on space radiation risk focuses on the following:
  - *Study type*:
    - *in vivo* studies involve living organisms, such as animals or humans, and provide direct insights into physiological responses;

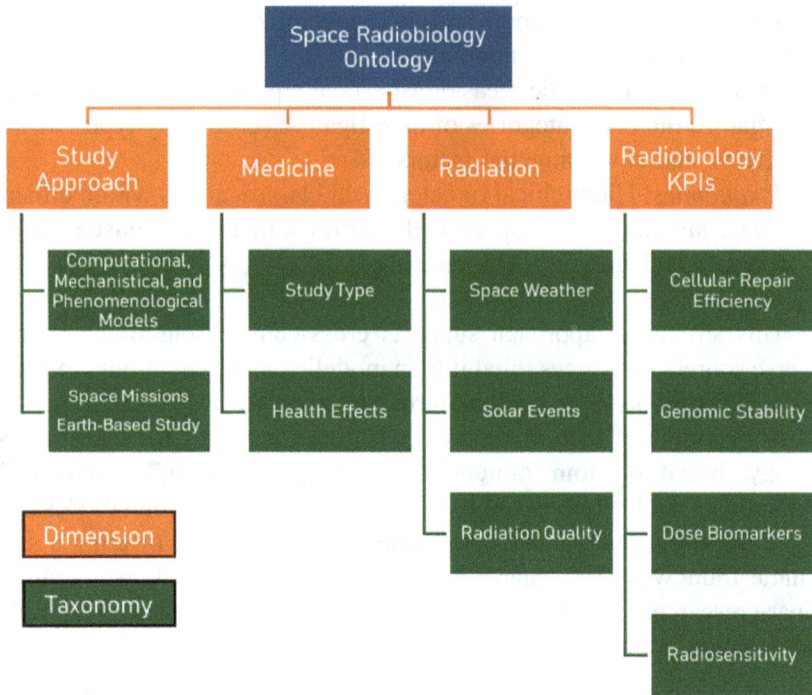

Fig. 6.1. A possible ontology for space radiobiology based on 11 taxonomies or dimensions (green boxes). This ontology was proposed [1] to categorize manuscripts for literature reviews and score them as relevant to the topic.

Source: Generated with licensed MS Copilot tool by the authors.

- *in vitro* studies using isolated cells or tissues, offering a controlled environment to study specific biological mechanisms;
- *in silico* approaches rely on computational simulations, bridging experimental findings with theoretical predictions.
- *Health effects*: Space radiation can lead to acute and chronic health effects, including cancer, cardiovascular disease, immune dysfunction, and neurodegeneration. Understanding these effects is crucial for developing countermeasures to protect astronauts on long-duration missions.

- *Radiation*: The unique environmental conditions of space drive space radiation risk:
  - *Space weather*: The variability of space weather, including solar activity and cosmic ray flux, affects radiation exposure levels.

Monitoring and predicting space weather events is critical for mission planning.

- o *Solar events*: SPEs, such as solar flares and coronal mass ejections, can result in intense bursts of high-energy particles, posing acute risks to the health of the astronaut.
- o *Radiation quality*: Unlike terrestrial radiation, space radiation consists of high-energy particles, including GCRs and heavy ions, which are more biologically *effective* and pose *significantly greater* risks.
- *Radiobiology KPIs*: To quantify and predict radiation effects, several biological metrics are used:
  - o *Cellular repair efficiency*: Cells' ability to repair DNA damage is a critical determinant of radiation sensitivity and long-term health risks.
  - o *Genomic stability*: Radiation can disrupt genomic integrity, leading to mutations and chromosomal aberrations, which underpin many radiation-induced diseases.
  - o *Dose biomarkers*: Biological markers, such as gene expression or protein levels, help measure radiation exposure and predict health outcomes.
  - o *Radiosensitivity*: Individual differences in susceptibility to radiation, influenced by genetic and physiological factors, are a key consideration.

Let's use an architectural analogy to make the ontology, dimensions, and taxonomies more accessible. Imagine you are designing a building. You would need a clear and organized approach to ensure that every part of the structure is well thought-out, functional, and cohesive. In this analogy, the building represents the overall knowledge structure used to understand and assess IR risks in space exploration.

The dimensions of the building can be compared to the foundation of a structure. Just as a foundation is a broad, essential element that supports the entire building, dimensions are the broad, overarching categories that support the overall ontology. These dimensions define the primary areas of study that help organize the risk assessment. In our case, dimensions include *Study Approaches, Medicine, Radiation, and Radiobiology KPIs*.

Each foundation (or dimension) provides the structural support necessary for building a clear, reliable understanding of radiation risks. The dimensions are not the details but rather the broad categories that hold the different parts of the study together. For example, Study Approaches (such as computational, mechanistic, and phenomenological models) form the fundamental base of the knowledge structure for understanding how we approach and model radiation risk.

Once you have a solid foundation, you need pillars or columns to provide vertical support, divide the space, and organize it into functional areas. In our analogy, the taxonomies are like the pillars. Taxonomies are the detailed classifications within each dimension, breaking down broad categories into more specific and manageable elements.

For example, under the Study Approaches foundation, you could have several pillars, such as Computational Models, Mechanistic Models, and Phenomenological Models. Each represents a specific area of research or a specific method used to assess radiation risks. Similarly, within the Medicine foundation, taxonomies might include specific Study Types (such as *in vivo*, *in vitro*, or *in silico*) and Health Effects (such as acute or chronic effects). These taxonomies are specific classifications that hold up and further organize each dimension, ensuring the overall structure is more precise and functional.

In any building, the elements inside (such as doors, windows, rooms, and materials) fill the space between the pillars and provide functional and aesthetic purposes. In our analogy, these elements can be compared to the dictionary of expressions.

In the context of risk assessment for space radiation, the dictionary of expressions refers to specific data, terms, processes, or even variables used within each taxonomy to provide further detail and clarity. These elements "fill" the dimensions and taxonomies, providing concrete, practical data to describe, measure, and assess radiation risks.

For example:

- In the Study Approaches dimension, the dictionary of expressions might include terms such as dose–response relationships, radiation interactions, or biological outcomes.
- In the Radiobiology KPIs dimension, specific biomarkers (such as γ-H2AX foci or DNA damage markers) could be examples of elements in the dictionary that help define or quantify radiation effects.

Just as a well-designed building requires solid foundations (dimensions), supportive pillars (taxonomies), and functional elements (dictionary of expressions), the study of IR risks in space needs a structured approach that consists of the following:

- Dimensions provide the broad, essential categories (foundations) for organizing research.
- Taxonomies break down these broad categories into specific, detailed areas of study (pillars).
- The dictionary of expressions fills each area with precise data, terminology, and processes (the materials and elements that make a building functional).

The risk assessment of IR in space exploration is a multifaceted challenge that integrates taxonomies, diverse study approaches, medical perspectives, radiation characteristics, and radiobiology KPIs. This comprehensive framework enables the identification of risks, the prediction of health outcomes, and the development of protective measures, ensuring the safety and well-being of astronauts on long-duration missions. Other approaches going deep, for example, into managing experimental datasets, constitute a base for knowledge foundation to be considered a must for effective field research.

Another relevant example of the use of ontology in the space radiobiology field can be found in the NASA RBO project, developed at NASA Ames Research Laboratory, whose aim was to create a knowledge infrastructure for managing large datasets of experimental data in this field of research [2].

6.3 Methods of Biological Damage Evaluation for Risk Assessments

Effective risk assessment methods are essential for understanding and mitigating the health risks associated with space radiation. This section explores the various techniques and methodologies used to evaluate these risks, providing a comprehensive overview of the field's current state.

6.3.1 *Dose estimation and relative biological effectiveness*

The concept of dose in radiation physics quantifies the amount of energy imparted by IR to a unit mass of matter. It is a foundational measure in assessing radiation effects on biological tissues [3]. Different types of dose concepts are introduced in dosimetry. Absorbed dose, a purely physical measure, does not account for biological variations and is used in dosimetry and material analysis. Equivalent dose introduces the concept of radiation quality to reflect the type of radiation's biological impact. Effective dose provides a single metric to evaluate overall risk, combining radiation type and tissue sensitivity. Primarily, it is used for radiation protection guidelines. Table 6.1 lists the types of doses used in radiation protection.

6.3.1.1 *Absorbed dose*

The absorbed dose, denoted as D, is defined as the energy deposited per unit mass and is expressed in gray (Gy), where

$$1 \text{ Gy} = 1 \text{ J/Kg}$$

While the absorbed dose provides a physical measure of energy deposition, it does not account for the varying biological effects caused by different types of radiation. For example, a dose of 1 Gy from low-linear energy transfer (LET) radiation (e.g., X-rays) may have significantly different biological consequences compared to a 1 Gy dose from high-LET radiation (e.g., alpha particles or heavy ions). To address this variability, two additional dose measures are introduced, as follows.

6.3.1.2 *Equivalent dose*

The equivalent dose, denoted as H_T, adjusts the absorbed dose to reflect the radiation's RBE. It is calculated as

$$H_T = \sum_R w_R \cdot D_{T,R}$$

where:

- $D_{T,R}$ is the absorbed dose in a tissue or organ T from radiation type R,
- w_R is the radiation weighting factor, representing the relative biological impact of radiation type R. Table 6.2 lists w_R values for different types of radiation, as defined in International Commission on Radiological Protection (ICRP) Publication No. 103 (2007).
  The equivalent dose is expressed in **sieverts (Sv)**, where

$$1 \text{ Sv} = 1 \text{ J/Kg}$$

Table 6.1. Characteristics of dose types in radiation protection.

Dose Type	Definition	Unit	Purpose	Key Equation
Absorbed (D)	Energy is deposited per unit mass of matter by ionizing radiation.	Gray (Gy)	Measures the physical energy imparted by radiation to matter.	$$D = \frac{\Delta E}{\Delta m}$$ where ΔE is energy deposited and Δm is the mass.
Equivalent (H_T)	Adjusts the absorbed dose to account for the biological effectiveness of different radiation types.	Sievert (Sv)	Evaluates the biological risk of radiation exposure in a specific tissue or organ.	$$H_T = \sum_R w_R \cdot D_{T,R}$$ where w_R is the radiation weighting factor.
Effective (E)	Aggregates equivalent doses across all tissues, weighted by tissue-specific radiation sensitivity factors.	Sievert (Sv)	Provides an overall measure of the biological risk to the entire organism.	$$E = \sum_T w_T \cdot H_T$$ where w_T is the tissue weighting factor.

Table 6.2. ICRP radiation weighting factors (w_R) for different types of radiation, as defined in ICRP Publication No. 103 (2007).

Radiation Type	ICRP Weighting Factor (w_R)	Rationale
X-Rays, Gamma Rays	1	Reference radiation with minimal ionization clustering and a baseline biological effect.
Beta Particles	1	Similar to gamma rays in terms of biological effectiveness.
Neutrons	5–20	Variable w_R depends on neutron energy; higher weighting is needed for intermediate energies.
Alpha Particles	20	High-LET radiation with dense ionization and a significantly higher biological impact.
Heavy Ions	20	Treated like alpha particles due to comparable LET and biological damage.

6.3.1.3 *Effective dose*

The effective dose, denoted E, accounts for the varying sensitivities of different tissues to radiation; it aggregates the equivalent doses across all tissues, weighted by tissue-specific dimensionless factors (w_T):

$$E = \sum_T w_T \cdot H_T$$

Like the equivalent dose, the effective dose is expressed in sieverts and provides a holistic measure of the organism's overall biological risk.

6.3.1.4 *Radiation biological effectiveness*

RBE [4, 5] measures the biological damage caused by radiation relative to a reference, typically X-rays or gamma rays, with an RBE of 1. Space radiation, dominated by high-LET particles such as heavy ions, often has a much higher RBE due to its dense ionization tracks, leading to more severe and complex biological damage. The RBE depends on factors such as radiation type, LET, dose, dose rate, and biological endpoints (e.g., cell survival, DNA damage, or cancer risk).

The ICRP recommends using RBE in terrestrial radiation protection. To simplify radiation weighting for practical applications, the ICRP introduces radiation weighting factors (w_R), which are standardized RBE-like values. These factors convert the physical dose (Gy) into an equivalent dose (Sv) that reflects the biological impact [6].

The RBE often exceeds terrestrial weighting factors for space radiation, especially for high-energy heavy ions. However, the ICRP framework provides a foundational approach for assessing mixed radiation fields in terrestrial and low-LET environments.

In space, the biological effects of high-LET radiation necessitate RBE values higher than ICRP recommendations for terrestrial settings. While terrestrial applications focus on low-LET radiation risks, space missions must consider additional factors, such as the enhanced cancer risk due to the high RBE of GCR heavy ions, acute effects like ARS during SPEs, and long-term tissue degeneration and CNS impacts.

Incorporating RBE into computational models and shielding designs enables mission planners to mitigate risks effectively, aligning with ICRP principles and space-specific requirements. Table 6.3 contains examples of exposition scenarios and radiation-correlated RBE estimations [4–10]. Each row outlines the radiation type, LET, biological endpoint, estimated RBE, and implications for astronauts and mission design.

These factors account for the varying biological effectiveness of different radiation types, enabling the calculation of equivalent doses in Sv for radiation protection.

6.3.2 *Exposure limits and NASA astronaut risk models*

NASA has long recognized the significant risks that space radiation poses to human health, which has led to the continuous evolution of standards and recommendations for radiation protection in space exploration. A key metric used in defining these standards is the concept of risk of exposure-induced death (REID), which represents the probability that radiation exposure will lead to fatal cancer or other detrimental health outcomes. NASA's guidelines are anchored around a maximum REID threshold of

Table 6.3.   Examples of space radiation exposure scenarios with their associated estimated RBE.

Scenario	Radiation Type	Typical LET (keV/µm)	Biological Endpoint	Est. RBE	Implications
GCR-Induced DNA Damage in Astronauts	Heavy ions (e.g., Fe)	~100–200	DNA double-strand breaks	10–20	Heavy ions create dense ionization tracks, leading to complex, irreparable DNA damage. This poses a significant risk for cancer and tissue degeneration.
SPE Exposure During a Solar Flare	Protons	~1–5	Acute radiation syndrome (ARS)	1–3	High fluence of SPE protons can cause ARS in unshielded astronauts, with symptoms including nausea, fatigue, and potential mortality if exposure is severe.
CNS Effects from GCRs	Mixed high-LET particles	~50–150	Neuronal damage and inflammation	5–10	GCR exposure is linked to cognitive decline and neuroinflammation due to the enhanced RBE of high-LET ions affecting brain tissue.
Long-Term Cancer Risks in Orbit	Protons and electrons	~0.2–2	Mutations and carcinogenesis	1–2	Chronic exposure to low-LET radiation in trapped belts contributes to cumulative cancer risk but has lower RBE than heavy ions.
Spacecraft Shielding Failure	Neutrons (secondary)	~10	Cell death and organ damage	3–5	Secondary neutrons generated by primary GCR interactions with spacecraft materials exacerbate tissue damage due to their moderate RBE.

3% throughout an astronaut's career, reflecting an acceptable balance between mission risks and long-term health considerations [11, 12].

Since the beginning of the space program, NASA's permissible exposure limits (PELs) have changed in response to advancements in radiation biology, risk modeling, and medical understanding:

- 1958–1983: Initial limits allowed up to 4 Sv exposure for males, with no defined standard for females.
- 1983–1989: The limit for both males and females was set at 2 Sv, reflecting new data on the doubling dose for heritable genetic effects.
- 1989–2000: Standards were revised to 2.5 Sv for males and 1.75 Sv for females, recognizing gender-based differences in radiation sensitivity.
- 2000–2003: Exposure limits were further reduced to 1.0 Sv for males and 0.6 Sv for females, emphasizing enhanced risk assessments.
- 2003–2022: During this period, limits dropped to approximately 0.3 Sv for males and 0.2 Sv for females, driven by concerns about long-term health impacts, particularly cancer risk.
- 2022 onward: The guidelines have been updated to a unified limit of 0.6 Sv for both males and females, reflecting parity in risk modeling and health considerations for astronauts of all genders.

The adjustments over time illustrate NASA's commitment to safeguarding astronauts' health through stringent radiation safety protocols, particularly for missions in low Earth orbit (LEO). By linking the standards to REID, NASA ensures that health risks are consistently evaluated and minimized, thus facilitating safer human space exploration.

6.3.3 *Track structure model*

The track structure model analysis is a key approach for understanding the spatial distribution of energy deposition by IR at the microscopic level. This is especially relevant for high-LET radiation, such as the heavy ions encountered in space. These particles generate dense ionization tracks that differ substantially from low-LET radiation's more uniform ionization patterns. A critical aspect of this analysis is the Katz model, which provides a quantitative framework for assessing radiation effects, and the concept of the penumbra track, which describes the spatial spread of secondary ionizations around the primary particle track [13, 14].

The Katz model quantifies the likelihood of biological effects, such as cell inactivation or DNA damage, caused by IR. It describes the radial dose distribution around the track of a charged particle (CP), emphasizing the localized damage caused by the particle's energy deposition.

The probability of a biological outcome is expressed as

$$P(x) = 1 - e^{-\sigma_o S(x)}$$

where:

- $P(x)$ is the probability of a specific effect at dose x;
- σ_0 is the inactivation cross-section, representing the target area sensitive to radiation;
- $S(x)$ is the cell survival probability (CSP) function derived from deposited energy, which is discussed in the following section.

The radial dose $D(r)$ around the particle's path is defined as

$$D(r) = \frac{LET}{2\pi r^2}$$

where:

- LET is the linear energy transfer of the particle,
- r is the radial distance from the particle's trajectory.

In addition to the central ionization track described by the Katz model, a *penumbral track* surrounds the core path. This penumbra is formed by secondary electrons, also known as delta rays, ejected by the high-energy CP as it traverses the medium. These electrons carry significant energy and can travel micrometers to millimeters from the primary track, creating a halo of additional ionizations.

The energy deposited in the penumbral region diminishes with distance from the track and is characterized by

$$D_{penumbra}(r) = krn$$

where:

- $D_{penumbra}(r)$ is the dose at distance r from the primary track;
- k is a proportionality constant depending on the particle type and energy;
- n is an exponent typically greater than 2, reflecting the steep energy falloff.

The penumbral track contributes significantly to radiation's biological effects by extending the potential damage region beyond the primary track. For high-LET particles, such as those found in GCRs, the penumbra intensifies the overall damage, especially in densely packed cellular environments like human tissue. This enhances radiation's RBE, as the damage encompasses direct hits to cellular nuclei and effects on nearby structures and molecules.

Combining core and penumbral tracks improves risk assessments for space radiation exposure. By incorporating these concepts into predictive models such as the Katz model, researchers can simulate the spatial distribution of energy deposition and its biological effects with high precision. This aids in designing shielding materials and medical countermeasures for astronauts. Figure 6.2 shows a representative track structure for a 5 MeV proton at the entrance region (a) and in the Bragg peak region (b), where maximum energy deposition occurs due to increased ionization density.

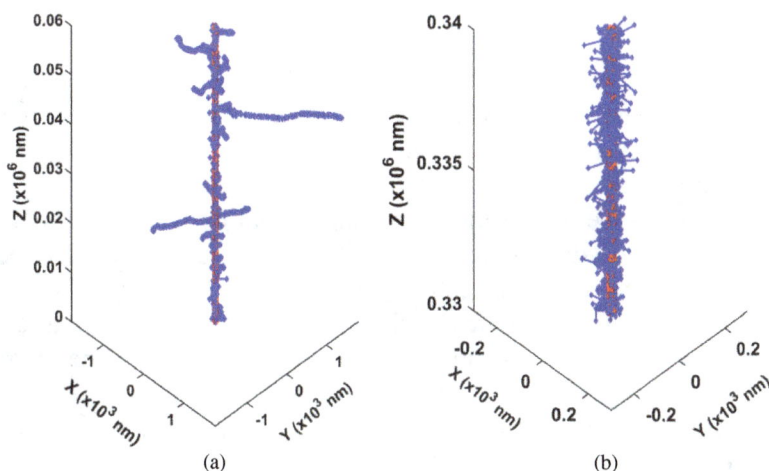

(a)　　　(b)

Fig. 6.2. A representative track structure for a 5 MeV proton at the entrance part (a) and in the Bragg peak region (b), described in Chapter 7. The proton was emitted along the positive Z direction. Red and blue dots represent the energy depositions by the proton and secondary electrons, respectively. Note: In the two subplots, we maintained the same aspect ratio for the z and x/y axes; however, we plotted them using different ranges.

Source: Youfang Lai *et al.*, Recent developments on gMicroMC: Transport simulations of proton and heavy ions and concurrent transport of radicals and DNA, June 2021 *International Journal of Molecular Sciences*, 22(12), 6615. CC BY 4.0 (https://creativecommons.org/licenses/by/4.0/).

6.3.4 *Cell survival probability functions*

Another important class of function that allows the modeling of the effects of IR on biological matter is the cell survival curves. CSP curves are crucial for understanding the biological effects of IR because they describe the relationship between the radiation dose absorbed by a population of cells and the fraction of cells that survive after irradiation. These curves are typically derived from clonogenic assays, which determine the number of colonies formed from a known number of irradiated cells, thus quantifying their ability to proliferate after radiation exposure [15].

The relevance of the cell survival curves to dose-effect models stems from their ability to provide insights into the direct biological damage caused by radiation, including DNA damage, mutations, and the activation of repair processes. As IR can produce various types of biological damage, from minor DNA damage to lethal double-strand breaks, the survival curves can reveal how well cells recover and how much damage is incurred at different radiation doses. Depending on their cell types and repair mechanisms, different tissues and organs will exhibit distinct survival curves, as cells in rapidly dividing tissues (e.g., bone marrow and the gut lining) tend to be more sensitive to radiation. In contrast, more differentiated cells (e.g., neurons) may be more resistant.

For different organs at risk, the cell survival curve varies due to factors such as tissue sensitivity, cell proliferation rates, and the presence of repair mechanisms. In critical tissues such as the central nervous system (CNS), cardiovascular system (CVD), and eye lenses (e.g., cataract formation), the dose–effect relationship is significant for understanding how radiation exposure during space missions could lead to delayed effects. For example, CNS cells may exhibit a unique survival curve, reflecting a lower turnover rate and limited ability to repair damage compared to tissue cells that regenerate quickly. Similarly, tissues such as bone marrow may show greater sensitivity to radiation, as these cells are frequently dividing and thus more vulnerable to DNA damage, leading to cancer or hematological disorders.

Several mathematical models have been developed to describe cell survival curves. These models range from more straightforward approaches, such as the linear-quadratic (LQ) model, to more complex ones that account for sublethal damage repair, track structure, and cellular heterogeneity. For instance, the LQ model is often used to describe the survival of rapidly dividing cells in tissues exposed to low-LET radiation. In contrast, more specialized models, such as the sublesion theory, account for the interactions of particles with biological tissue at the microscopic level, which is crucial for understanding the effects of heavy ions or high-LET radiation, as encountered in space environments.

These survival curves and associated models are essential in dose–effect relationship models used for radiation risk assessment, particularly in space radiation protection. By integrating survival curve data with cosmic radiation flux measurements, mathematical models can predict the potential biological outcomes of space radiation exposure, such as cancer induction, neurological damage, or cardiovascular effects. Such predictions are critical for assessing astronaut health during space missions, particularly for prolonged missions to Mars, where the risk of radiation exposure is significantly higher than in LEO.

The following is a list of the CSP models we used in our work, described in Chapter 8, to develop dose–effect models for space radiobiology. The variable D refers to the absorbed dose, while F is the flux.

6.3.4.1 *Target theory: n-target N-hit model (nTNH)*

Based on Marie Curie's pioneering work on radioactivity in the late 19th and early 20th centuries, Hugo Fricke formulated the target theory in the 1920s. It models how IR interacts with cells, proposing that critical cellular structures, named "target," must be "hit" to cause effects such as cell death. In this model, each target (e.g., DNA) is destroyed after receiving N hits, and the cell dies once all targets within it have been destroyed.

$$S(D) = 1 - (1 - B)^n$$

$$B = e^{\frac{-D}{D_0}} \left[1 + \sum_{2}^{N} \frac{\left(\frac{-D}{D_0} \right)^{N-1}}{(N-1)!} \right]$$

where:

- D_0 is the mean lethal dose for which the mean number of lethal events per cell equals 1 – at a dose $D0$, the fraction of cell survival is equal to $1/e$ or 37% (for the single-target single-hit model);

- n is the number of identical targets in the cell that are susceptible to being damaged from the IR;
- N is the number of hits necessary to destroy a single target.

Two special cases of nTNH include:

(a) single-target single-hit model (sTSH),
(b) single-target N-hit model (sTNH).

6.3.4.2 *Cellular track structure theory*

The cellular track structure theory (TST) is a radiobiological model used to describe the effects of CPs on biological systems. It is rooted in the multi-target single-hit model and was first introduced by Robert Katz in 1968. TST defines the action cross-section as the probability that radiation will activate sensitive targets (e.g., cell structures), leading to a measurable biological endpoint, such as cell death.

In this model, radiation-induced damage arises from two primary sources: direct ionization by the charged particle (ion) itself and delta rays, or secondary electrons, produced as the ion passes through biological matter.

The model assumes that the cell survival probability (S) results from the combined effects of these two damage mechanisms. Specifically, it is expressed as the product of the probabilities of survival from ion-kill (Π_i) and gamma-kill (Π_y) components:

$$S = \Pi_i \times \Pi_y$$

This formulation allows the theory to differentiate between densely ionizing events (associated with ion-kill) and more diffuse energy deposition (associated with gamma-kill), providing a nuanced understanding of radiation effects at the cellular level [13].

The gamma-kill probability is calculated by

$$\Pi_v = [1 - (1 - e^{D_o})]$$

The ion-kill probability is given by

$$\Pi_i = e^{-(\Sigma_0 P_S(Z,E)F)}$$

where

$$P_s(Z_{CP}, E_{CP}) = \left[1 - e^{\frac{-Z^{*2}}{k\beta_c^2}}\right]^{m_S}$$

in which multi-target detectors, such as cells, are represented in TST by the following four parameters:

- m_s is the number of targets per cell,
- D_0 is the radiosensitivity,
- Σ_0 is the cross-section saturation value,
- k is the detector saturation index,

while:

- Z_{CP} is the charge number of the particle,
- E_{CP} is the kinetic energy of the particle,
- Z^* is the effective charge number of the particle,
- β_c is the particle velocity relative to the velocity of light.

Considering the track structure model equations, we can express the CSP model for CP as

$$S(D) = \Pi_v * \Pi_i$$

6.3.4.3 *Linear-quadratic model*

In 1972, Kellerer and Rossi introduced the linear-quadratic (LQ) model, in which a lethal event is supposed to be caused by a single hit due to one particle track (the linear component αD) or due to the accumulated damage due to two particle tracks (the quadratic component βD^2).

$$S(D) = e^{-\alpha D - \beta D^2}$$

where α and β are the linear and quadratic coefficients with dose induction terms.

6.3.4.4 *Linear-quadratic model modified by hyper-radiosensitivity effect*

The LQ model is a monotonically decreasing function of dose; it cannot be used to describe low-dose phenomena of hyper-radiosensitivity (HRS)/ increased radio resistance (IRR). To account for such phenomena, in 1990, Joiner, in collaboration with colleagues, developed a modified version of the LQ model, called induced repair (IR), which is given by

$$S(D) = e^{[-\alpha(1+(\frac{\alpha_s}{\alpha}-1)e^{\frac{-D}{D_c}})D-\beta D^2]}$$

where:

- D_c is the dose at which the transition from HRS to IRR occurs,
- $\alpha_s > \alpha$ is the initial slope of the surviving fraction curve at $D = 0$ Gy and represents the IRR at low doses.

6.3.4.5 *Linear-quadratic-cubic model*

This model was introduced by Tobias in 1985 to better describe the IR effects at the high doses usually used in radiotherapy treatments by adding a cubic term to the polynomial function of the LQ model:

$$S(D) = e^{-\alpha D-\beta D^2+\gamma D^3}$$

where:

- αD is the linear component,
- βD^2 is the quadratic component,
- γ is the cubic coefficient with dose induction terms.

6.3.4.6 *Sublesion theory: repair-misrepair model*

Proposed by Tobias in 1985, the repair-misrepair (RMR) model describes the evolution of the function $U(t)$, which reflects the mean number of lesions before any repair activation. The yield of the initially induced lesions, U_0, was considered proportional to the dose D:

$$U_0(D) = \delta D$$

where δ is the proportionality constant of the radiation quality.

By considering that the linear repair is not a perfect process and that the repair time is limited in time, the survival equation becomes

$$S_\phi(D) = e^{-U_0(D)} \left[1 + \frac{U_0(D)(1-e^{-\lambda t_r})}{\epsilon} \right]^{\epsilon \phi}$$

where:

- t_r is the repair time after irradiation,
- λ is the rate constant for linear repair processes,
- ϵ is the ratio of λ and K,
- ϕ is the probability that self-repair steps are perfect eurepairs (or good repair).

6.3.4.7 Sublesion theory: lethal potentially lethal model

Curtis developed, in 1986, the *lethal-potentially lethal* model, which considers the repair process. He proposed a classification of the radio-induced lesions:

- lesions that are unrepairable and are, therefore, lethal;
- potentially lethal lesions for which the repair process is activated.

The survival equation that allows us to predict the survival ratio is

$$S(D) = e^{-N_{tot}(D)} \left[1 + \frac{N_{PL}(D)}{\epsilon(1-e^{-\epsilon_{PL}t_r})} \right]^{\epsilon}$$

where:

- T is the irradiation time,
- t_r is the available repair time after irradiation,
- $N_{tot}(D)$ is the number of total lesions (sum of lethal lesions and potentially lethal lesions) at the end of the exposure time T,
- $N_{PL}(D)$ is the number of potentially lethal lesions at the end of the exposure time T,
- ϵ_{PL} the constant per unit of time repair rate,

- ϵ_{2PL} the constant per unit of the time rate of interaction between two potentially lethal lesions,
- ϵ is the ratio between ϵ_{PL} and ϵ_{2PL}.

6.3.4.8 *Sublesion theory: saturable repair model*

In 1985, Goodhead proposed a model, called the saturable repair model, based on the hypothesis that the efficiency of the repair system decreases with the dose and that this decrease is caused by the saturation of the repair kinetics [15]. Using this hypothesis, the survival equation becomes

$$S(D) = e^{-\frac{n_0(D)-C_0}{1-\frac{C_0}{n_0(D)}kt_r(C_0-n_0(D))}}$$

where:

- t_r is the time available for repair after irradiation,
- $n_0(D)$ is the initial number of lesions due to dose D,
- C_0 is the initial number of available repair molecules or enzymes,
- k is the proportionality coefficient,

Table 6.4 summarizes the proposed CSP models, along with the year of introduction and the leading applications of each. In the table, each model's suitability for these applications is rated on a scale ranging from one * to five *, where five * indicates the highest relevance. Including these application ratings highlights the diverse contexts in which these models are utilized, providing a comparative view of their practical utility. The "Other" column in the table refers to broader applications that are not strictly limited to the fields explicitly mentioned, such as radiation biology, radiotherapy, low-dose radiotherapy, space radiation, or heavy nuclei studies. These may include applications such as studying radiation effects in materials science, agriculture (e.g., plant biology), food preservation, modeling population-level exposure scenarios (e.g., nuclear accidents), or serving as foundational tools in basic research and education on radiation interactions. This column assigns stars based on each model's general adaptability and relevance to these less conventional applications.

Table 6.4. Categorized cell survival models under ionizing radiation by their year of introduction and application across various fields such as radiation biology, radiotherapy, low-dose radiotherapy, space radiation, and heavy nuclei.

Year	Model	Radio-biology	Radio-therapy	Low-Dose Radio-therapy	Space Radiation	Heavy Nuclei	Other
1920s	Target	*****	****	***	***	***	**
1960s	TST	*****	**	**	****	*****	***
1970s	LQ	*****	*****	****	***	***	***
	S-LPL	*****	****	***	****	***	***
1980s	LQC	****	****	***	***	***	***
1990s	S-LE-QM	*****	****	***	****	****	***
	S-LE	****	***	***	****	****	***
	LQ-HRS	****	****	*****	***	***	***
	S-SR-D	****	****	***	****	****	****
	S-SR-RT	****	****	***	****	****	****

6.3.5 *Monte Carlo computational simulations*

MC simulations are a class of computational techniques that use random sampling to solve problems that might be deterministic but are too complex for analytical solutions. They are particularly effective in modeling stochastic processes, such as the interaction of radiation with matter, making them indispensable for studying space radiation and its biological effects. The MC method was developed during the 1940s as part of the Manhattan Project [16], with physicist Stanislaw Ulam and mathematician John von Neumann playing key roles in its creation. Ulam conceived the idea while considering the probabilistic nature of card games, which inspired him to use random sampling for complex mathematical problems. This approach was named "Monte Carlo" as a nod to the famous casino, emphasizing the method's reliance on randomness and probability. Since then, MC methods have been widely applied in various fields, including physics, engineering, finance, and medical sciences.

In the context of radiation transport and dosimetry, MC techniques simulate the paths of individual particles, accounting for their interactions with matter at every step. This level of detail allows for highly accurate predictions of dose distributions, energy deposition, and secondary particle production, all of which are critical for space radiation studies.

MC-based computational simulations are a cornerstone of modern space radiation risk assessment. They offer detailed insights into the interactions of radiation with biological tissue and materials. These simulations rely on stochastic methods to model the complex physical processes governing radiation transport and interactions, making them indispensable for predicting radiation dose distributions and assessing the effectiveness of shielding.

Several advanced MC codes are widely used in this domain, including GEANT4, FLUKA, HETC-HEDS, MCNP6, and PHITS, each offering specialized functionalities:

- GEANT4 [17, 18] is a highly versatile toolkit for simulating the passage of particles through matter, covering electromagnetic and hadronic interactions. It is particularly notable for the GEANT4-DNA library, which allows simulations at molecular and cellular scales. This extension is invaluable for studying radiation-induced biological damage, such as DNA strand breaks and chromosomal aberrations, providing critical insights into the microdosimetric effects of space radiation.
- *FLUKA* [19] is known for comprehensively modeling high-energy particle transport and nuclear interactions. It is frequently used for spacecraft shielding analysis and dosimetry, ensuring adequate protection against GCRs and SPEs. FLUKA's precision in simulating mixed radiation fields makes it a key tool for space applications.
- *HETC-HEDS* (High-Energy Transport Code for Human Exploration and Development in Space) [20] specializes in modeling high-energy nuclear interactions and transport tailored for space exploration scenarios. It is adept at simulating the secondary radiation produced when primary space radiation interacts with shielding or human tissue.
- *MCNP6* (Monte Carlo N-Particle Transport Code) [21] is renowned for its versatility in modeling neutron, photon, and CP transport. Its applications in space radiation research include dose calculations, shielding optimization, and neutron-induced secondary radiation studies, which are critical for ensuring astronaut safety during long-duration missions.
- *PHITS* (Particle and Heavy Ions Transport System) [22] focuses on space-specific radiation environments, integrating solar and cosmic radiation spectra for precise simulations of dose and risk. Its adaptability to mission-specific conditions makes it a valuable asset in planning and risk assessment.

MC simulations enhance the understanding of dose distributions in human tissues and spacecraft materials under various conditions. These tools integrate biological models with physical data, enabling comprehensive assessments of health risks and the development of protective strategies. Their synergy with experimental data ensures robust and reliable predictions for mission planning [22].

6.4 Experimental Studies in Space Radiobiology

This section focuses on experimental research methodologies for studying space radiation's biological effects. It emphasizes how experiments in complementing computational and theoretical models by providing real-world data to validate and refine radiation risk predictions.

The basic techniques involved in both *in vitro* and *in vivo* experimentation in space will be described. Additionally, the use of IR sources in ground-based facilities, primarily particle accelerators that simulate space radiation environments, will be discussed.

6.4.1 *Biological models and radiation exposure in space radiobiology*

Biological models and experimental setups are critical for studying the effects of space radiation on living organisms. These experiments provide valuable insights into radiation-induced biological damage and the associated risks for human space exploration. This section outlines commonly used biological models, radiation exposure protocols, and advanced platforms for conducting experiments in space and on Earth.

6.4.1.1 *Biological models*

Space radiobiology research employs a variety of biological models, each tailored to address specific scientific questions related to the effects of space radiation. Mammalian cell lines, such as fibroblasts and cancer cells, are foundational in controlled studies of DNA damage and cellular repair mechanisms, allowing researchers to investigate radiation-induced apoptosis under defined laboratory conditions. Human stem cells, particularly pluripotent and hematopoietic stem cells, are invaluable for exploring tissue regeneration and differentiation, offering insights into how radiation affects the body's ability to repair itself.

Rodent models, including mice and rats, serve as comprehensive systems for studying the systemic impacts of space radiation. These models, enhanced by transgenic technologies, allow organ-specific investigations and mimic many aspects of human biology. Yeast cells, such as *Saccharomyces cerevisiae*, are pivotal in space missions such as BioSentinel due to their simple genome and utility in studying DNA repair and metabolic responses to radiation.

The fruit fly, *Drosophila melanogaster*, provides a rapid-cycling genetic model to understand mutagenesis and hereditary effects, while zebrafish, with their transparent embryos, facilitate developmental studies and visualization of radiation's effects on organogenesis. Similarly, the nematode *Caenorhabditis elegans (C. elegans)* is a robust model for understanding neurodegenerative processes and cellular stress responses due to its simplicity, genetic tractability, and high reproducibility. Advanced models such as 3D organoids, which are grown from stem cells to mimic tissue architecture, allow researchers to study organ-specific radiation responses in a way that mirrors human biology. Plants such as *Arabidopsis thaliana* add another dimension by offering a perspective on how radiation affects cellular structures such as chloroplasts, relevant for long-term space habitat sustainability. Finally, humanized mouse models, engineered to express human genes or immune systems, closely replicate human physiological responses, making them indispensable for studying immune function and cancer development under space radiation exposure.

This diverse array of biological models, summarized in Table 6.5, highlights the field's integrative approach. The table highlights their unique features, specific applications, and advantages in investigating the effects of space radiation and microgravity on living systems. These models range from mammalian cell lines and human stem cells to whole-animal systems such as rodents, providing insights across molecular, cellular, and systemic levels.

6.4.1.2 *Radiation exposure protocols*

In space radiobiology, radiation exposure protocols are designed to mimic the diverse and complex conditions that organisms encounter in space. These protocols are vital for understanding the risks posed by space radiation and devising protective measures. Acute exposure, where a high dose of radiation is delivered over a short period, is significant for studying the effects of intense radiation bursts, such as those caused by SPEs.

Table 6.5. Summary of biological models used in space radiobiology.

Biological Model	Description	Strong Points
Mammalian Cell Lines	Cultured human and animal cells, e.g., fibroblasts, epithelial cells, and cancer cell lines.	Controlled environment; detailed study of DNA damage, repair, and apoptosis.
Human Stem Cells	Includes pluripotent stem cells and hematopoietic stem cells.	Models for tissue regeneration are sensitive to radiation and ideal for studying differentiation effects.
Rodent Models	Mice and rats are used for whole-organism studies.	Systemic response evaluation; transgenic models available; replicates human biology at organ levels.
Yeast Cells	*Saccharomyces cerevisiae* are often used in CubeSat missions like BioSentinel.	Simple genome; insights into DNA damage/repair mechanisms; rapid growth cycle.
Drosophila (Fruit Fly)	Tiny insect with a well-mapped genome.	Short lifecycle; excellent for genetic studies and radiation-induced mutagenesis research.
Zebrafish	Transparent embryos are used for developmental studies.	Enables visualization of developmental effects and genetic similarity to humans.
C. elegans (Nematode Worm)	Tiny, transparent nematode with a simple nervous system.	Highly reproducible results; model for neurodegeneration and stress responses.
3D Organoids	Miniaturized versions of organs grown from stem cells.	Mimics tissue architecture; allows complex studies of organ-specific responses to radiation.
Plant Models	*Arabidopsis thaliana* and other plants.	Studies effects on cellular structures like chloroplasts; insights into ecological implications.
Humanized Mouse Models	Rodents are engineered to express human genes or immune systems.	Closely mimics human physiological responses, ideal for cancer and immune system studies.

Such events can deliver significant radiation doses over hours or days, posing acute risks to astronauts. Conversely, chronic exposure protocols focus on replicating the continuous, lower-dose radiation environment characteristic of GCRs during long-term space missions. These studies are crucial for understanding cumulative damage and the body's ability to adapt or repair over extended periods.

Another key protocol involves fractionated exposure, involving doses administered in segments over time. This approach helps researchers study cumulative effects while observing potential recovery mechanisms between exposures, providing insights into how biological systems manage intermittent radiation stress.

Mixed-field radiation protocols are designed to simulate the intricate radiation environments of space, combining particles of varying types and energies. This complexity reflects the true nature of space radiation, enhancing the realism of experimental conditions. Additionally, variable dose rates explore the impact of different exposure speeds, shedding light on how biological responses vary with radiation intensity.

Lastly, protocols that integrate simulated microgravity with radiation exposure are becoming increasingly relevant. These combined studies examine how the absence of gravity and radiation interact to influence biological responses, as both factors play significant roles in the health risks associated with space travel.

These protocols collectively ensure a comprehensive understanding of space radiation's biological effects, providing a foundation for developing effective countermeasures and supporting the safety of future space exploration.

6.4.1.3 *Facilities for biological experiments in space*

The International Space Station (ISS) is an unparalleled microgravity and space radiation environment for conducting biological experiments that would be impossible to perform on Earth. Within this environment, specialized facilities such as the Bioculture System and *KUBIK* support advanced cell culture experiments. These bioreactors are designed to grow and sustain cell cultures in microgravity while enabling detailed investigations into how space radiation affects gene expression, DNA repair mechanisms, and cellular metabolism. By offering a controlled environment for studying cellular behavior in space, these platforms provide critical insights into the fundamental biological processes influenced by the absence of gravity and exposure to cosmic radiation.

Instruments installed on "phantoms" and sophisticated dosimeters monitor radiation exposure in real-time to ensure precision in radiation experiments aboard the ISS. These tools provide detailed data on radiation levels, creating a robust framework for assessing its effects on biological systems. This combination of cutting-edge bioreactors and radiation-sensitive payloads makes the ISS a vital laboratory for advancing space radiobiology.

Complementing ISS-based studies, *CubeSat* [23] has emerged as a transformative platform for space radiobiology research. These small, modular satellites, typically ranging from 1U to 12U, provide a cost-effective means of deploying biological payloads into LEO and beyond. Equipped with microfluidic systems, radiation sensors, and biological samples, CubeSats enable experiments to probe the effects of space radiation on living organisms.

Notable missions such as NASA's *BioSentinel* [23] utilize yeast cells to investigate DNA damage and repair in deep space, marking a significant leap beyond low Earth orbit studies. *BioSentinel* incorporates fluidic cards and compact optical systems to monitor biological responses during flight, pushing the boundaries of autonomous space-based research.

CubeSats also hold immense potential for supporting future lunar and interplanetary missions. Their adaptability allows them to function as biological payloads for lunar landers and orbiters, facilitating studies in environments that simulate extraterrestrial conditions. These missions will be instrumental in understanding the combined effects of space radiation and reduced gravity, offering critical data to protect human explorers venturing into deep space.

Integrating ISS facilities with CubeSat technology creates a synergistic approach to space radiobiology, expanding the scope of research across diverse platforms. This combination paves the way for a deeper understanding of radiation's risks and biological resilience, ultimately supporting humanity's ambition to explore and inhabit space.

6.4.2 *Accelerator infrastructure and specialized equipment for testing*

As mentioned in Section 5.5.2.2, most experimental research in space radiobiology relies heavily on advanced accelerator infrastructure. A particle accelerator is a device that uses variable electromagnetic fields to accelerate CPs, such as protons or heavy ions, and guides them along a controlled trajectory in the form of organized beams or particle bunches. These beams are then directed onto target samples, enabling researchers to study the behavior of biological systems, materials, and electronics under irradiation. The main characteristics defining a particle accelerator are the type of particle or isotope it can accelerate and the maximum achievable kinetic energy. These attributes determine the relevance of an accelerator for simulating the space radiation environment, which consists of a mix of high-energy protons, alpha particles, and heavy ions from GCRs and SPEs.

The two primary accelerator architectures used in radiobiology are linear accelerators and circular colliders.

- *Linear accelerator (LINAC)*: In LINACs, particles are accelerated in a straight line through radiofrequency cavities. Each cavity applies a synchronized electric field, progressively boosting the particles' energy. LINACs are well suited for experiments requiring a single, highly focused beam with minimal energy spread. A LINAC example is the Heavy Ion Medical Accelerator in Chiba (HIMAC), which uses linear acceleration to produce monoenergetic ion beams, ideal for cellular and tissue-level radiation studies.
- *Circular colliders*: Using powerful magnets to bend particle paths, circular colliders accelerate particles in a closed-loop trajectory. These accelerators achieve higher energy levels due to multiple passes through the same acceleration components. However, CPs in circular colliders experience energy loss due to synchrotron radiation, which becomes significant for lighter particles such as electrons. For example, the GSI Helmholtz Centre employs a circular collider synchrotron to deliver high-energy heavy ions that mimic cosmic rays.

The choice between linear and circular designs depends on the type of particles required, the energy range, and the experimental objectives. Table 6.6 presents a list of advanced accelerator facilities, including NASA's Space Radiation Laboratory (NSRL) also described in Fig. 6.3, GSI's Heavy Ion Research Facility, and Europe's FAIR; their role in mimicking space radiation, particularly through the generation of high-energy protons, heavy ions, and mixed-radiation fields, is well established and emphasized. The table highlights key characteristics such as the country of operation, primary isotopes/particles used, energy ranges, special features, and references for further reading.

Table 6.6. Comparison of significant accelerator facilities worldwide.

Facility Name	Primary Isotopes Particles	Energy Ranges (MeV/u)	Special Features
NASA GCR Simulator (NSRL) [24], USA	P, He, C, Si, Fe	50–1,000	Simulates GCR spectrum, monoenergetic, and mixed radiation fields.
GSI Helmholtz Centre [25], Germany	P, He, C, Si, Fe	50–1,000	Advanced ion beam therapy and space radiation studies; pioneer in heavy-ion radiobiology.
FAIR@GSI (Facility for Antiproton and Ion Research) [25], Germany	Heavy ions (P to U)	Up to 2,000	High-intensity beams for studying heavy ions under extreme conditions.
J-PARC (Japan Proton Accelerator Research Complex) [26], Japan	P, neutrons	Protons: 400–1,000	Focuses on neutron and proton radiation effects, with applications in materials and biology.
HIMAC (Heavy Ion Medical Accelerator) [27], Japan	Carbon, oxygen	50–800	Combines medical research with heavy-ion space radiobiology studies.
GANIL (Grand Accélérateur National d'Ions Lourds) [28], France	Heavy ions	50–600	Flexible heavy-ion beams for biological and material sciences.
CERN (European Organization for Nuclear Research) [29], Switzerland	P, lead ions	Protons: up to 7,000 GeV	Studies ultrahigh-energy cosmic rays and particle interactions relevant to astrophysics.
HIRFL (Heavy Ion Research Facility in Lanzhou) [30], China	P, C, Ni, N, Fe	50–1,200	Studies heavy-ion physics and radiobiological effects of high-LET radiation.
INFN-LNL (Istituto Nazionale di Fisica Nucleare, Laboratori Nazionali di Legnaro) [31], Italy	P, He, heavy ions	30–200	Focuses on low-energy particle interaction studies for biological applications.
CNAO (National Centre for Oncological Hadrontherapy) [32], Italy	Carbon, protons	50–400	Primarily a clinical center but also conducting radiobiology and particle therapy research.

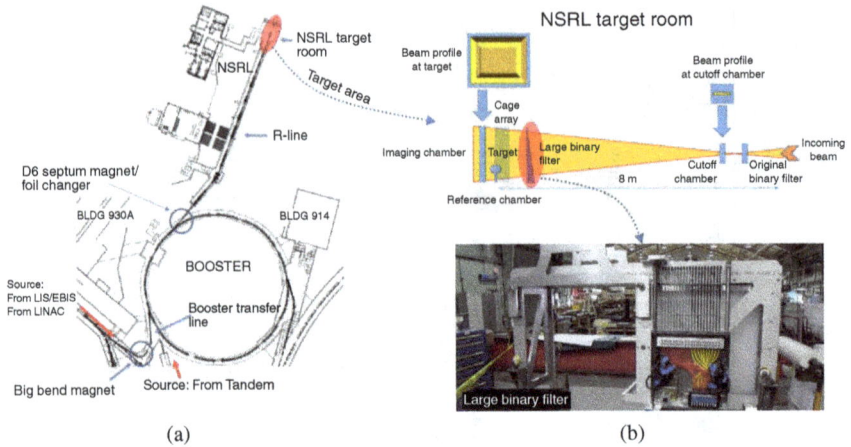

Fig. 6.3. Facility layout of NSRL at BNL: (a) Tools to reliably control system hardware settings from ion production by the LIS through booster injection, acceleration, extraction, and delivery to the NSRL target room were developed to deliver the GCR simulator ion-energy beam combinations sequentially. (b) Position of imaging chamber behind target (top, left-hand side), cut-off chamber (top, right-hand side) near beam entrance to target room, and photo of large-area degrader (binary filter) system (bottom) in the NSRL beam line to maintain control and uniformity of a 60×60 cm^2 beam. BNL, Brookhaven National Laboratory; EBIS, electron beam ion source; GCR, galactic cosmic radiation; LINAC, linear accelerator; LIS, laser ion source; NSRL, NASA Space Radiation Laboratory.

Source: Simonsen, L. C., *et al.* (2020). NASA's first ground-based galactic cosmic ray simulator: Enabling a new era in space radiobiology research. *PLoS Biol* 18(5): e3000669. CC0 (https://creativecommons.org/public-domain/cc0/).

In addition to the primary accelerator, a range of complementary equipment is essential for practical experimentation:

- *Dosimetry systems*: These measure the delivered radiation dose with high precision. They are critical for ensuring experimental consistency and correlating biological effects with dose levels. Systems include ionization chambers, thermoluminescent dosimeters (TLDs), and solid-state detectors.
- *Microbeam irradiators*: These devices deliver highly localized radiation doses to specific areas, such as single cells or subcellular components. This allows for studying radiation effects with high spatial precision.

- *Detectors for monitoring radiation*: Detectors, such as scintillators, proportional counters, and silicon detectors, monitor radiation fields during experiments. They provide real-time feedback on particle fluence, energy spectra, and dose distribution.
- *High-resolution imaging systems*: For biological studies, imaging systems such as confocal and electron microscopes are integrated to analyze radiation-induced damage at the cellular and molecular levels.
- *Target sample holders and environmental chambers*: During irradiation experiments, these maintain biological samples under controlled conditions, such as temperature and humidity.

Figure 6.3 represents the NASA Space Radiation Laboratory, one of the most relevant particle accelerator-based facilities for mimicking space radiation. In contrast, Fig. 6.4 gives an idea of how a test plan can be structured using such infrastructure so that the complexity of space radiation environments can be effectively emulated at the end of the irradiation.

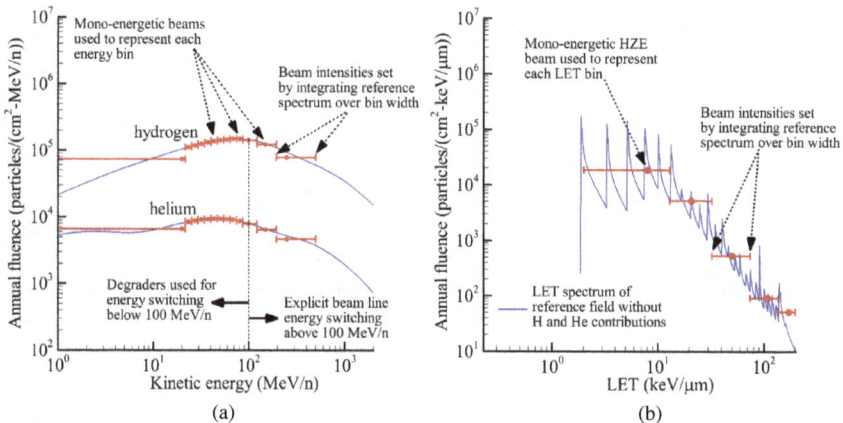

Fig. 6.4.   Representation of the reference field using discrete monoenergetic beams. The hydrogen and helium energy spectra are considered directly (a), whereas HZE ions are represented within the LET spectrum (b). Solid blue lines are the reference spectra. The bin widths for 1 GeV/n protons and helium particles are at lower fluences and not shown in the figure. HZE, high charge and high energy ions; LET, linear energy transfer.

Source: Simonsen, L. C., Slaba, T. C., Guida, P., Rusek, A. (2020). NASA's first ground-based Galactic Cosmic Ray Simulator: Enabling a new era in space radiobiology research. *PLoS Biol* 18(5): e3000669. https://doi.org/10.1371/journal.pbio.3000669. This is an open access article, free of all copyright, and may be freely reproduced, distributed, transmitted, modified, built upon, or otherwise used by anyone for any lawful purpose. The work is made available under the Creative Commons CC0 public domain dedication.

References

[1] Bartoloni, A., *et al.* (2024). Synergistic advances in space radiation health effects: Collaborative insights from AMS Roma Sapienza and Medical Physics Division of IRCCS University Hospital of Bologna Hospital. *Proceedings of the 75th International Astronautical Congress (IAC)*, Milan, Italy.

[2] Berrios, D., *et al.* (2022). The radiation biology ontology: A new tool supporting FAIR principles across radiation biology facilitating data discovery and integration radiobiology with heavy charged particles: A historical review. *Conference Proceedings of EPWR2022.*

[3] ICRU. (1998). *Fundamental Quantities and Units for Ionizing Radiation.* ICRU Report 60.

[4] Suzuki, M., Kase, Y., Yamaguchi, H., Kanai, T., and Ando, K. (2000). Relative biological effectiveness for cell-killing effect on various human cell lines irradiated with heavy-ion medical accelerator in Chiba (HIMAC) carbon-ion beams. *International Journal of Radiation Oncology Biology Physics*, 48(1), 241–250. https://doi.org/10.1016/s0360-3016(00)00568-x.

[5] Hunter, N., and Muirhead, C. R. (2009). Review of relative biological effectiveness dependence on linear energy transfer for low-LET radiations. *Journal of Radiological Protection*, 29(1), 5–21. https://doi.org/10.1088/0952-4746/29/1/R01.

[6] ICRP. (2007). *The 2007 Recommendations of the International Commission on Radiological Protection.* ICRP Publication 103.

[7] Durante, M., and Cucinotta, F. A. (2011). Physical and biological space radiation effects. *Reviews of Modern Physics*, 83(4), 1245–1281.

[8] Cucinotta, F. A., and Durante, M. (2006). Cancer risk from exposure to galactic cosmic rays: Implications for space exploration by humans. *PLoS Medicine*, 3(8), e331.

[9] Zeitlin, C., *et al.* (2013). Measurements of energetic particle radiation in transit to Mars on the Mars Science Laboratory. *Science*, 340(6136), 1080–1084.

[10] Parihar, V. K., *et al.* (2016). Cosmic radiation exposure and persistent cognitive dysfunction. *Scientific Reports*, 6, 34774.

[11] National Council on Radiation Protection and Measurements (NCRP). (2006). *Information Needed to Make Radiation Protection Recommendations for Space Missions Beyond Low-Earth Orbit.* NCRP Report No. 153.

[12] Norbury, J. W., *et al.* (2016). Galactic cosmic ray simulation at the NASA Space Radiation Laboratory. *Life Sciences in Space Research*, 8, 38–51.

[13] Waligórski, M. P., Grzanka, L., and Korcyl, M. (2015). The principles of Katz's cellular track structure radiobiological model. *Radiation Protection Dosimetry*, 166(1-4), 49–55. https://doi.org/10.1093/rpd/ncv201.

[14] Hill, M. A. (1999). Radiation damage to DNA: The importance of track structure. *Radiation Measurements*, 31(1-6), 15–23. https://doi.org/10.1016/s1350-4487(99)00090-6.

[15] Bodgi, L., *et al.* (2016). Mathematical models of radiation action on living cells: From the target theory to the modern approaches. A historical and critical review. *Journal of Theoretical Biology*, 394, 93–101. https://doi.org/10.1016/j.jtbi.2016.01.018.

[16] Metropolis, N. (1987). The beginning of the Monte Carlo method. *Los Alamos Science*, 15, 125–130.

[17] Agostinelli, S., *et al.* (2003). GEANT4: A simulation toolkit. *Nuclear Instruments and Methods in Physics Research Section A*, 506(3), 250–303.

[18] Bernal, M. A., *et al.* (2015). Track structure modeling in radiation biology: GEANT4-DNA extensions. *Physics in Medicine & Biology*, 60(14), R1–R63.

[19] Battistoni, G., *et al.* (2016). Overview of the FLUKA code. *Annals of Nuclear Energy*, 82, 10–18.

[20] Werneth, C. M., *et al.* (2020). Development and application of HETC-HEDS for space radiation transport. *Life Sciences in Space Research*, 25, 45–54.

[21] Goorley, T., *et al.* (2013). *MCNP6 User's Manual Version 1.0*. Los Alamos National Laboratory Report LA-CP-13-00634.

[22] Zaman, F. A., Townsend, L. W., de Wet, W. C., Looper, M. D., Brittingham, J. M., Burahmah, N. T., Spence, H. E., Schwadron, N. A., and Smith, S. S. (2022). Modeling the lunar radiation environment: A comparison among FLUKA, Geant4, HETC-HEDS, MCNP6, and PHITS. *AGU*. https://doi.org/10.1029/2021SW002895.

[23] Harandi, B., Ng, S., Liddell, L. C., Gentry, D. M., and Santa Maria, S. R. (2022). Fluidic-based instruments for Space Biology Research in CubeSats. *Frontiers in Space Technologies*, 3. https://doi.org/10.3389/frspt.2022.853980.

[24] NSRL website. https://www.bnl.gov/nsrl/ (last accessed 12 December 2024).

[25] GSI website. https://www.gsi.de/en/start/news (last accessed 12 December 2024).

[26] J-PARC website. https://j-parc.jp/c/en/ (last accessed 12 December 2024).

[27] HIMAC website. https://www.qst.go.jp/site/qst-english/ (last accessed 7th August 2025).

[28] GANIL website. https://www.ganil-spiral2.eu/ (last accessed 12 December 2024).

[29] CERN website. https://home.cern/ (last accessed 12 December 2024).

[30] HIRFL website. https://english.imp.cas.cn/research/facilities/HIRFL/ (last accessed 7th August 2025).

[31] INFN-LNL website. https://www.lnl.infn.it/ (last accessed 12 December 2024).

[32] CNAO website. https://fondazionecnao.it/ (last accessed 12 December 2024).

Chapter 7

Ionizing Radiation Medical Applications

7.1 Introduction

Ionizing radiation plays a pivotal role in modern medicine, offering invaluable tools for both diagnostic and therapeutic applications. It is characterized by its ability to deposit energy into matter, resulting in the ionization of atoms and molecules. In diagnostics, ionizing radiation enables the visualization of organs and tissues with millimetric precision. At the same time, in therapy, it is harnessed to target and destroy diseased tissues, such as in cancer treatment. The versatility of ionizing radiation in medicine stems from its capacity to be finely controlled and adapted to diverse clinical needs.

While ionizing radiation offers remarkable benefits, its interaction with biological systems can produce significant effects. These effects range from ionizing molecules within cells to the potential induction of cellular damage, which may lead to tissue injury or longer-term consequences such as carcinogenesis. The biological impact of radiation depends on factors such as the type of radiation, the absorbed dose, and the exposure duration. At low doses, typical in diagnostic imaging, the risk is minimal and well-managed through established safety protocols. At higher doses used in therapeutic applications, the goal is to maximize the destruction of tumoral cells while minimizing harm to surrounding healthy tissue. Thus, continuous optimization is observed in both research and clinical practice to optimize diagnostic protocols (i.e., improving image quality while reducing diagnostic doses) and deliver higher doses to the tumors while sparing non-target tissues. This process involves clinicians and medical physicists performing measurements and adopting or developing adequate metrics.

In the following sections, we explore specific examples of its use in diagnostics and therapy, illustrating various modalities employed, the rate at which radiation is delivered (dose rate), and the absorbed dose associated with each type of examination or treatment to illustrate the careful balance between efficacy and safety in the medical use of ionizing radiation.

7.2 Diagnostic Applications of Ionizing Radiation

Radiological equipment using ionizing radiation is crucial in medical diagnostics. It offers detailed insights into organs and tissues, including their functions. The evolution of medical radiological imaging technology and its wide availability have resulted in an exponential increase in utilization.

Ionizing radiation includes X-rays and gamma rays, which have enough energy to ionize atoms and molecules, potentially causing cellular damage. Evidence for a risk of cancer arising from radiation doses of less than 100 mSv remains limited [1]. However, when used judiciously, these radiological techniques provide invaluable diagnostic information. Patient dose optimization is needed.

This chapter explores various types of radiological and nuclear medicine equipment that utilize ionizing radiation for diagnostic purposes, including X-ray systems with image intensifiers, computed tomography (CT) scanners, single photon emission computed tomography (SPECT), and positron emission tomography (PET). The absorbed dose range for a typical diagnostic examination is provided for each type of equipment.

7.2.1 *X-ray systems with image intensifiers*

X-ray systems with image intensifiers are commonly used in diagnostic radiology, particularly in fluoroscopy. These systems enhance the visibility of internal structures by converting X-rays into visible light, which is then amplified for real-time viewing.

An X-ray beam passes through the patient and strikes the image intensifier, which converts the X-rays into visible light. This light is then amplified and projected onto a monitor, allowing for continuous observation of dynamic processes, such as gastrointestinal studies, angiography, and orthopedic procedures.

Fluoroscopy has seen significant advancements in recent years, particularly in image quality, dose optimization, and functionality. Modern fluoroscopic systems are equipped with flat-panel detectors (FPDs), which offer superior spatial resolution and reduced noise, compared to traditional image intensifiers. These detectors improve visualization during barium swallows, cardiac catheterizations, and orthopedic surgeries, ensuring greater diagnostic accuracy and procedural success.

Another notable development is the implementation of advanced dose-reduction technologies, such as pulsed fluoroscopy and adaptive filtering (e.g., directional adaptive filter kernels and a ridgeness filter), which minimize radiation exposure to patients and clinicians without compromising image quality [2–3].

Furthermore, real-time image processing and integration with artificial intelligence (AI) enable automatic adjustments of parameters such as brightness and contrast, enhancing workflow efficiency [4].

Portable fluoroscopy systems have become more compact and maneuverable, making them invaluable in operating rooms and emergency settings. Advances in angiography have been driven by the need for high-resolution imaging, reduced radiation exposure, and improved visualization of blood vessels. Modern angiographic systems employ digital subtraction angiography (DSA), which removes background structures in real time, providing more precise and detailed images of vascular abnormalities. Integrating three-dimensional (3D) imaging and rotational angiography allows clinicians to reconstruct complex vascular structures, thereby improving diagnostic precision and procedural planning (e.g., for radioembolization) [5].

Additionally, using lower radiation dose protocols and developing contrast agents with improved biocompatibility have enhanced patient safety.

In orthopedic surgery, fluoroscopic guidance has been transformed by innovations in navigation and imaging systems. The introduction of intraoperative 3D fluoroscopy and cone-beam CT provides surgeons with highly detailed, real-time images, enabling the precise placement of hardware, such as screws, plates, and prosthetics. These systems also allow for immediate verification of implant positioning, reducing the need for follow-up corrections and improving surgical outcomes.

Robotic-assisted systems have further revolutionized orthopedic procedures by integrating fluoroscopic imaging with navigation software to enhance the accuracy of surgical interventions. AI algorithms now assist in identifying anatomical landmarks and suggesting optimal hardware placement paths, saving time and reducing errors. Furthermore, dosage management systems in modern fluoroscopy reduce radiation exposure for patients and surgical teams, ensuring safer practices in the operating room. These advancements collectively enhance the precision, safety, and efficiency of orthopedic surgeries, improving patient outcomes.

The absorbed dose from fluoroscopy varies widely depending on the procedure's complexity and duration. Typical absorbed doses range from 1 to 50 milligrays (mGy) per procedure.

7.2.2 *Mammography*

Mammography has been a cornerstone of breast cancer screening and diagnosis for decades, evolving significantly to improve accuracy and reduce patient discomfort. Early systems relied on analog film-based imaging, which provided adequate sensitivity but limited dynamic range and image processing capabilities. The shift to digital mammography marked a transformative step, offering enhanced image quality, lower radiation doses, and improved workflow efficiency. Digital systems allow for post-acquisition image manipulation, such as magnification and contrast adjustment, aiding in the detection of subtle abnormalities.

A significant innovation in mammography is digital breast tomosynthesis (DBT), commonly known as 3D mammography [6]. DBT acquires multiple images from different angles, reconstructing a 3D representation of the breast. This reduces the impact of overlapping tissues, a limitation in traditional 2D mammography, and improves cancer detection rates, particularly for women with dense breast tissue. Further advancements include the integration of AI, which aids radiologists in identifying potential lesions by analyzing mammographic images with high precision. AI algorithms are particularly effective at detecting microcalcifications and masses, potentially reducing false positives and radiologists' workload. Additionally, contrast-enhanced mammography combines standard mammography with contrast agents to highlight vascular activity, aiding in the evaluation of suspicious lesions.

These advancements demonstrate continuous innovation in mammography, enhancing its role in early breast cancer detection while improving patient outcomes and diagnostic efficiency.

7.2.3 *Computed tomography*

CT scanners combine X-rays with computer processing to produce cross-sectional images of the body. They provide more detailed information than standard X-rays. CT scanners use a rotating X-ray tube and a series of detectors around the patient. As the X-ray tube rotates, multiple images are captured from different angles. A computer then reconstructs these images into detailed, cross-sectional views.

AI also revolutionizes CT reconstruction by enhancing image quality, reducing noise, and accelerating processing times. Traditional methods, such as filtered back projection (FBP) and iterative reconstruction, are computationally intensive and may compromise image resolution at lower radiation doses. AI-driven techniques, such as deep learning-based reconstruction, utilize neural networks trained on high-quality datasets to generate improved (i.e., denoised) images from raw CT data. AI enables ultra-low-dose imaging while maintaining diagnostic accuracy and addressing patient safety concerns associated with radiation exposure [7–9]. Adaptive algorithms can also correct for motion artifacts, enhancing precision in challenging scenarios. As AI integration progresses, it promises to transform CT imaging by optimizing accuracy, safety, and efficiency.

Spectral CT, or dual-energy or multi-energy CT, represents a groundbreaking advancement in CT technology [10]. Unlike conventional CT scanners that utilize a single energy spectrum for imaging, spectral CT leverages two or more distinct energy levels to differentiate materials based on their unique attenuation profiles. This technology improves tissue characterization, enhances contrast resolution, and reduces artifacts, such as beam hardening. Spectral CT has shown great promise in various clinical applications, including the precise identification of kidney stones, better visualization of bone marrow abnormalities, and improved detection of vascular diseases. Additionally, generating virtual monoenergetic images and iodine maps enhances diagnostic confidence while potentially reducing the quantity of contrast agents. The impact of spectral CT extends beyond diagnostics. In oncology, for instance, it aids in distinguishing between tumor tissue and surrounding structures, facilitating treatment

Table 7.1.   CT application, purpose, and typical absorbed doses.

Application	Purpose	Typical Absorbed Dose Range (mGy)
Head CT	Detects tumors, hemorrhages, and brain injuries.	30–50
Chest CT	Evaluates lung conditions, including cancer, infections, and pulmonary embolisms.	5–20
Abdominal CT	Diagnoses abdominal pain, kidney stones, and bowel obstructions.	10–30
Cardiac CT	Assesses coronary artery disease.	10–30

planning. Moreover, the technology supports lower radiation doses by optimizing image acquisition, aligning with the growing emphasis on patient safety. As spectral CT becomes more widely adopted, its role in advancing personalized medicine continues to expand, offering a powerful tool for diagnosis and follow-up in various diseases. Table 7.1 summarizes typical CT applications and correlated patient absorbed dose in mGy.

7.2.4 *Single photon emission computed tomography*

SPECT is a nuclear medicine imaging technique that uses gamma rays to create detailed functional imaging of organs and tumors.

SPECT involves injecting a gamma-emitting radioisotope into the patient's bloodstream. The radioisotope accumulates in the target tissues, emitting gamma rays that are detected by a gamma camera rotating around the patient. A computer processes these signals to construct cross-sectional images.

It can be used to evaluate normal physiology and monitor many diseases, including cardiovascular diseases [11–14], disorders of the central nervous system [15–17], cancer [18–21], and brain functions [22].

A key innovation of this technique has been hybrid imaging, which integrates SPECT with other modalities, such as CT (SPECT/CT). This combination enhances anatomical localization by fusing functional imaging from SPECT with structural imaging from CT. SPECT/CT is particularly valuable in oncology for accurate tumor staging and in orthopedics for pinpointing sources of pain or inflammation. Additionally, advancements in reconstruction algorithms, including iterative reconstruction techniques, have reduced noise and artifacts, enabling the use of lower tracer doses while maintaining diagnostic accuracy. These developments have significantly expanded the utility of SPECT in personalized medicine and precision diagnostics.

More recently, SPECT has undergone significant technological evolution, improving its diagnostic accuracy, sensitivity, and versatility in clinical applications. One significant advancement is the development of solid-state detectors, such as the cadmium-zinc-telluride (CZT) technology, which offers superior energy resolution compared to conventional sodium iodide (NaI) detectors. These detectors enable higher sensitivity and sharper image quality, allowing for earlier and more precise detection of diseases in areas such as cardiology, oncology, and neurology. CZT-based SPECT systems also support faster acquisition times, reducing patient burden and improving workflow efficiency in clinical environments.

AI has also played a transformative role in SPECT technology [23–25].

AI-based algorithms enable image reconstruction, segmentation, and quantification, improving diagnostic confidence and reducing the time required for image interpretation. These tools can identify subtle abnormalities in imaging data, aiding clinicians in the early detection of conditions such as coronary artery disease and neurodegenerative disorders. AI-driven advancements have also enhanced motion correction, addressing challenges posed by patient movement during imaging and resulting in higher-quality scans.

Dynamic SPECT imaging represents another cutting-edge innovation, enabling the capture of real-time physiological processes, such as blood flow and tracer kinetics. This capability is particularly valuable in cardiac imaging, as it allows for precise assessment of myocardial perfusion, and in neuroimaging, it provides insights into cerebral blood flow and receptor binding. By combining dynamic imaging with quantitative data analysis, clinicians can better understand disease mechanisms, advancing diagnostics and therapeutic monitoring. These advancements continue to solidify SPECT's role as a critical tool in modern nuclear medicine.

The evolution of portable devices enables the use of complementary SPECT/CT information to monitor the transit and accumulation of radionuclide activity during radionuclide theaphy [26]. Table 7.2 summarizes typical SPECT applications and correlated patient absorbed dose in mGy.

Table 7.2.   SPECT application, purpose, and typical absorbed doses.

Application	Purpose	Typical Absorbed Dose Range (mGy)
Cardiac SPECT	Evaluate myocardial perfusion and detect coronary artery disease.	7–15
Brain SPECT	Assess cerebral blood flow and detect neurological disorders.	5–10
Bone SPECT	Identify bone metastases and fractures.	4–88

7.2.4.1 *Type of radiopharmaceuticals*

SPECT utilizes several key radionuclides to produce detailed images of the body's internal structures and functions. The most commonly used radionuclide in SPECT is technetium-99m (Tc-99m), which is favored for its ideal physical properties, including a short half-life of about six hours and gamma emission of 140 keV, making it suitable for a variety of diagnostic applications. Tc-99m is used in several radiopharmaceuticals, such as Tc-99m methylene diphosphonate (MDP) for bone scintigraphy to detect fractures and bone tumors, Tc-99m Sestamibi for myocardial perfusion imaging to assess blood flow to the heart muscle, and Tc-99m hexamethylpropyleneamine oxime (HMPAO) for brain perfusion imaging to evaluate cerebral blood flow and detect conditions like stroke. Other Tc-99m-based radiopharmaceuticals include Tc-99m dimercaptosuccinic acid (DMSA) for renal cortical imaging and Tc-99m macroaggregated albumin (MAA) for lung perfusion imaging, which helps detect pulmonary embolism.

Another important radionuclide in SPECT is Iodine-123 (I-123), which emits gamma rays with an energy of 159 keV and has a half-life of approximately 13 hours. I-123 is primarily used for thyroid imaging with I-123 sodium iodide, helping to evaluate thyroid function and detect nodules and cancer. Additionally, I-123 ioflupane (DaTSCAN) is employed for brain imaging to assess dopamine transporter levels, aiding in the diagnosis of Parkinson's disease and other movement disorders.

Lastly, indium-111 (In-111), which emits gamma rays with energies of 171 and 245 keV and has a half-life of approximately 2.8 days, is used for specialized imaging. In-111 oxyquinoline (oxine) is employed to label white blood cells for detecting infection and inflammation, while In-111 pentetreotide binds to somatostatin receptors on neuroendocrine tumors, facilitating their localization and evaluation.

In summary, SPECT imaging relies on various radionuclides and their associated radiopharmaceuticals to provide detailed diagnostic information. Tc-99m is the most versatile and widely used, with applications spanning bone, cardiac, brain, renal, and pulmonary imaging. I-123 is particularly useful for thyroid and brain imaging, while Tl-201 is primarily used for cardiac assessments. Ga-67 helps detect infections and tumors, and In-111 is valuable for specialized imaging of infections and neuroendocrine tumors. Each radionuclide's unique properties, including half-life and gamma emission energy, make them suitable for specific diagnostic tasks, enabling clinicians to obtain precise and critical insights into various medical conditions. PET/CT scans now replace many SPECT/CT investigations with appropriate radionuclides.

7.2.5 *Positron emission tomography*

PET is a nuclear medicine imaging technique that provides high-resolution images of metabolic processes in the body. Widely used in oncology, cardiology, and neurology, PET relies on injecting a positron-emitting radioisotope, such as fluorine-18, which accumulates in the target tissue. Positrons interact with electrons, producing gamma photons that are detected by the scanner to construct detailed metabolic activity images.

PET/CT technology advancements, particularly systems with an extended long axial field of view (LAFOV), have significantly enhanced molecular imaging [27]. Traditional PET/CT systems capture limited anatomical ranges per scan, often requiring multiple acquisitions for broader coverage (see Fig. 7.1(a)). In contrast, modern systems with total-body or larger FOV allow simultaneous imaging of extensive regions or the entire body (see Fig. 7.1(b)), improving sensitivity by up to 40 times and reducing acquisition times. These improvements are invaluable in oncology, enabling more efficient detection of metastases while enhancing patient comfort and workflow in clinical settings. LAFOV systems also support real-time tracking of tracer kinetics throughout the body (see Fig. 7.1(c)), which is critical for clinical and research applications. In oncology, this innovation facilitates faster and more comprehensive assessments, while in research, it enables detailed analysis of dynamic processes.

In parallel, silicon photomultipliers (SiPMs) have revolutionized PET/CT performance [28]. These advanced detectors enhance spatial

(a)

(b)

(c)

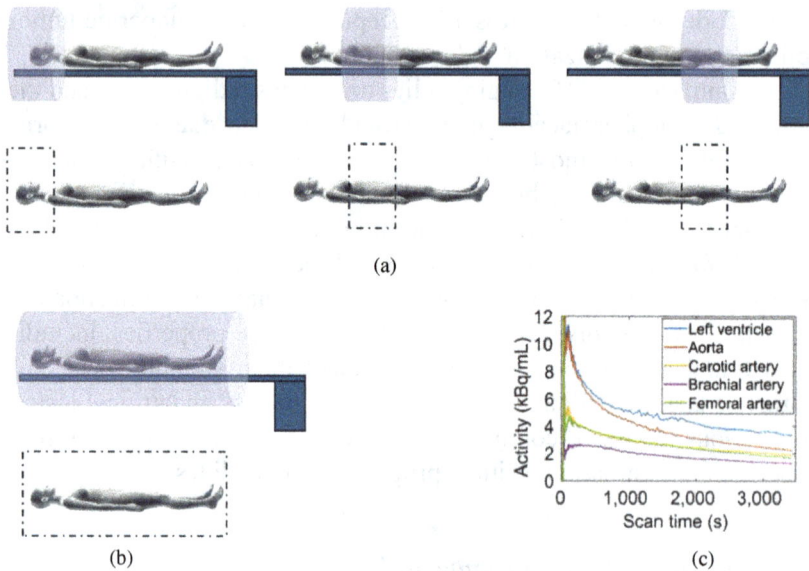

Fig. 7.1. PET diagnostic machine. (a) Conventional short-axis PET scanner, (b) Large axial FOV PET scanner, (c) Example of time-activity curves (TACs) of a radiopharmaceuticals such as [^{18}F]-FDG derived from the mean counts /or activity or SUV in a region of interest (i.e., ROI, representing an organ or a tumoral lesion) drawn on images collected over time using a whole-body dynamic PET/CT scanner.

Source: Generated with licensed MS PowerPoint and R script code by the authors.

resolution and enable time-of-flight (TOF) imaging, which precisely localizes positron annihilation events, thereby improving image clarity and lesion detectability even in patients with low tracer uptake or high body mass. Novel radiotracers expand PET/CT's applications, targeting specific cancers, neurodegenerative diseases, and inflammatory conditions. These innovations have solidified PET/CT's role in precision medicine, enabling tailored diagnostics and therapy monitoring.

AI has transformed PET/CT, optimizing data processing and interpretation [29, 30]. AI-powered algorithms enhance image reconstruction by reducing noise and improving resolution, enabling lower radiotracer doses and shorter scan times, which in turn minimizes radiation exposure while increasing efficiency. Machine learning models excel in detecting abnormalities, improving diagnostic accuracy by identifying subtle lesions, and reducing errors. AI-driven tools also streamline tasks such as tumor segmentation and tracer uptake quantification, accelerating metabolic activity assessments, and tumor volume analysis. In research, AI predicts disease progression and evaluates treatment response by analyzing imaging biomarkers, advancing personalized healthcare.

Dynamic imaging adds another dimension to PET/CT technology by capturing sequential images over time, offering valuable insights into physiological and metabolic processes. This approach is particularly impactful in cardiac imaging, where it quantifies myocardial blood flow, and in oncology, where it assesses tumor heterogeneity and treatment response. Combining temporal data with quantitative analysis enhances disease understanding and supports more informed clinical decisions.

Integrating LAFOV systems, dynamic imaging, advanced detectors, and AI has propelled PET/CT to the forefront of molecular imaging. These innovations improve precision diagnostics and patient outcomes; they also expand the modality's potential in research and personalized care, shaping the future of healthcare.

PET/magnetic resonance (MR) imaging represents a breakthrough in hybrid imaging, combining the molecular sensitivity of PET with the superior soft tissue contrast of MR. This integration has evolved through significant technological advancements, addressing challenges in hardware compatibility, image processing, and clinical application.

Initially, combining PET and MR presented substantial technical hurdles. PET systems rely on photodetectors that are sensitive to electromagnetic interference, while MR systems generate strong magnetic fields. Early attempts to integrate these modalities required innovative shielding techniques and the development of components that could operate without interference. The introduction of SiPMs marked a turning point in PET/MR systems. Unlike traditional photomultiplier tubes, SiPMs are compact, magnetic field resistant, and capable of delivering high-resolution images. This compatibility allowed the seamless integration of PET detectors into the MR gantry, enabling simultaneous data acquisition without compromising performance.

Advances in image reconstruction algorithms have played a crucial role in PET/MR's evolution. Hybrid imaging requires accurate attenuation correction to merge PET's functional data with MR's anatomical information. Early approaches relied on MR-based tissue segmentation, but recent developments incorporate AI-powered techniques to enhance accuracy

Table 7.3.   PET application, purpose, and typical absorbed doses.

Application	Purpose	Typical Absorbed Dose Range (mGy)
Oncology	Detect and monitor cancer and its metastases	7–14
Cardiology	Evaluate myocardial perfusion and viability	4–10
Neurology	Assess brain disorders such as Alzheimer's disease and epilepsy	7–14

and reduce artifacts. The main steps of the evolution of PET/MR technology are summarized by Li *et al.* (2024) [31]. Table 7.3 summarizes typical PET applications and correlated patient absorbed dose in mGy.

7.2.5.1 *Radionuclides and radiopharmaceuticals used in PET*

PET utilizes positron-emitting radionuclides to provide high-resolution images of metabolic processes in the body. Fluorine-18 (F-18), gallium-68 (Ga-68), and carbon-11 (C-11) are the most commonly used radionuclides. Each offers unique properties that make it suitable for various diagnostic applications, as described in the following:

- *Fluorine-18 (F-18)*: Due to its optimal physical properties, fluorine-18 is the most widely used radionuclide in PET imaging. These include a half-life of approximately 110 minutes and its ability to form strong bonds with carbon, making it suitable for a wide range of radiopharmaceuticals. Common radiopharmaceuticals comprise the F-18 fluorodeoxyglucose (FDG) used in oncology, cardiology, and neurology. FDG is a glucose analog that accumulates in cells with high metabolic activity. It is the workhorse of PET imaging, used to detect and monitor various cancers, assess myocardial viability, and evaluate brain disorders such as epilepsy and Alzheimer's disease. F-18 fluorodopa (F-DOPA) is used for neurological applications to assess dopamine synthesis in the brain, aiding in diagnosis and monitoring.

  F-18 fluorothymidine (FLT) is used in oncology to measure cellular proliferation. Its versatility in forming a variety of radiopharmaceuticals makes it indispensable in clinical diagnostics and research. F-18 allows centralized production and distribution to PET imaging centers, making it widely accessible for clinical use.

- *Gallium-68 (Ga-68)*: Gallium-68 is another important radionuclide used in PET imaging. It has a short half-life of approximately 68 minutes and is typically obtained from a germanium-68/gallium-68 generator, making it convenient for on-site production without needing a cyclotron.

    Common radiopharmaceuticals in the oncology field include Ga-68 DOTATATE. It binds to somatostatin receptors, which are overexpressed in neuroendocrine tumors. It provides high sensitivity and specificity for detecting and staging these tumors. Ga-68 PSMA-11 targets the prostate-specific membrane antigen (PSMA) found in high amounts in prostate cancer cells. It is highly effective for imaging prostate cancer, including primary staging and detecting metastases. The Ga-68 generator allows the use of Ga-68-labeled radiopharmaceuticals at facilities without a cyclotron. Its ability to form highly specfic compounds makes it particularly valuable for targeted oncology and infection diagnostics imaging.

- *C-11 Choline*: Carbon-11 has a short half-life of approximately 20 minutes, necessitating on-site production using a cyclotron. Despite this limitation, C-11 is a valuable radionuclide for PET imaging due to its ability to be incorporated into a wide range of biologically active molecules. This enables the study of various physiological and biochemical processes in the body. C-11 choline is mainly used for oncological applications, primarily for imaging prostate cancer and brain tumors. It is taken up by cells involved in membrane synthesis, which is increased in many tumors. C-11 methionine is used for neurological and oncological applications. It is utilized to image brain tumors and assess amino acid metabolism. It is particularly valuable in distinguishing between tumor recurrence and radiation necrosis. The short half-life of C-11 limits its use to facilities with an on-site cyclotron and radiochemistry capabilities. However, its versatility in labeling different molecules makes it invaluable for research and specialized clinical applications.

7.3 Radiation Therapy Applications Based on Particle Acceleration

Therapeutic radiology, also known as radiation therapy (RT), uses advanced devices to treat cancer and other conditions by delivering precisely controlled doses of radiation to targeted tissues. These devices maximize damage to diseased cells while sparing surrounding healthy tissue. Most radiotherapy devices rely on photons (X-rays) or particles, such as protons and heavier ions, generated using particle accelerators. Photon-based radiotherapy technologies, including conventional radiotherapy, intensity-modulated radiation therapy (IMRT), and stereotactic radiotherapy, remain the most commonly used methods in clinical practice due to their versatility and efficacy.

Meanwhile, particle therapy, which uses protons or ions, offers a more precise dose delivery, making it particularly effective for tumors near critical structures or in pediatric patients. This diverse arsenal of radiotherapy equipment allows for tailored treatment plans that align with each patient's specific needs, ensuring optimal therapeutic outcomes.

Each type of equipment, from Roentgen therapy machines to sophisticated MR-LINAC systems, provides unique advantages in targeting and treating tumors while sparing healthy tissues. The typical dose ranges for various treatment modalities, conventional, hypofractionated, and stereotactic, are designed to maximize therapeutic efficacy while minimizing side effects, highlighting the precision and versatility of modern radiotherapy techniques. Intraoperative radiation therapy (IORT) equipment offers a valuable adjunct to conventional radiotherapy techniques by delivering focused radiation directly to the tumor bed during surgery. Devices such as the Mobetron, Intrabeam System, and LIAC linac provide surgeons and radiation oncologists with options for delivering precise radiation doses while minimizing exposure to healthy tissues. Typical dose ranges for IORT vary depending on the specific device and treatment protocol, but they generally aim to provide effective tumor control while minimizing side effects. Incorporating IORT into cancer treatment protocols allows for personalized and targeted therapy, ultimately improving outcomes for patients undergoing surgical resection of tumors.

The technology, energy, and dose rate choice depend on tumor characteristics, location, and the patient's condition. Photon-based systems excel in mobility and superficial treatments, while electron-based IORT is more versatile for various tumor depths. Ongoing research into technologies (e.g., FLASH-RT) is expected to improve cancer treatment outcomes.

The evolution of radiation delivery techniques has enabled tailored treatment approaches that improve outcomes while minimizing toxicity. From conventional methods, such as 3D-CRT with wedges and field-in-field techniques, to frontier approaches, such as 4π radiotherapy LATTICE, GRID and FLASH RT, each modality addresses specific clinical needs. As research advances, integrating novel technologies with established methods promises to enhance RT's precision and efficacy.

This comprehensive overview covers a wide array of therapeutic radiology equipment, including Roentgen therapy machines, brachytherapy devices (both low- and high-dose-rates and pulsed-dose-rate applications), linear accelerators (LINACs) with electronic portal imaging devices (EPIDs) and cone beam computed tomography (CBCT), tomotherapy, CyberKnife, MR-based LINACs, CT-based LINACs, PET-based LINACs, proton therapy, and ion therapy devices.

Section 7.3.1 describes the more relevant devices used for RT, which is followed by the strategies for dose delivery to patients, distinguishing between the ones used for many years in hospitals for treatment (Section 7.3.2) and the most promising technique still in the development phase (Section 7.3.3).

7.3.1 *Devices for RT*

LINACs play a central role in modern RT, providing exact and adaptable treatment options for cancer patients. These devices generate high-energy electron and photon beams, enabling effective tumor targeting while minimizing damage to surrounding healthy tissues. Advanced imaging technologies integrated into LINACs, such as CBCT and EPID, further enhance treatment precision by verifying patient positioning in real time.

Innovative systems, such as MR-LINACs and PET-LINACs, combine imaging modalities with radiation delivery to provide personalized and adaptive care. Additionally, specialized techniques, including stereotactic treatments and IORT, expand the therapeutic capabilities of LINACs to address diverse clinical needs with precision and efficacy. This section explores the various LINAC-based technologies, their applications, and their impact on patient outcomes in radiation oncology.

7.3.1.1 *Linear accelerators (electrons and photons)*

LINACs are the most common equipment used in external beam radiotherapy. They accelerate electrons to produce high-energy X-rays or electron beams, which are used to target tumors. EPID-equipped LINACs use an electronic imaging device to verify the patient's position and alignment before and during treatment, thereby ensuring precise radiation delivery. CBCT-equipped LINACs provide 3D imaging to verify patient positioning and anatomy immediately before treatment, improving accuracy and outcomes.

Table 7.4.   Linac-based treatments and typical absorbed doses.

Machine	Treatment Approach	Typical Dose Range (Gy)
LINACs, Tomotherapy	Conventional	1.8–2.0 Gy per fraction, with 50–70 Gy doses
	Hypofractionated	2.5–4 Gy per fraction, with 30–40 Gy doses
	Stereotactic	5–20 Gy per fraction, total doses of 20–60 Gy over 1–5 fractions
CyberKnife	Conventional	Rarely used
	Hypofractionated	6–10 Gy per fraction
	Stereotactic	12–25 Gy per fraction, typically 1–5 fractions

Tomotherapy machines combine the principles of CT scanning with RT. The device delivers radiation in a helical (spiral) pattern, allowing for highly conformal and precise dose distribution.

CyberKnife is a robotic radiosurgery system that delivers highly precise, non-invasive treatments using real-time imaging and robotic guidance. It is especially effective for treating small, well-defined tumors.

MR-LINAC systems combine magnetic resonance imaging (MRI) with LINACs to provide superior soft tissue contrast for real-time imaging and adaptive radiotherapy. CT-based LINACs integrate imaging capabilities with LINACs to enhance treatment planning and delivery through accurate anatomical imaging. PET-LINAC systems combine positron emission tomography (PET) with linear accelerators to provide metabolic imaging, enhancing the targeting of active tumor regions. Table 7.4 summarizes typical LINAC and typical correlated patient absorbed dose ranges in Gy.

7.3.1.2 *Ion therapy devices*

Ion or heavy ion therapy uses charged particles such as carbon ions for treatment. These particles provide more precise energy deposition than protons or X-rays. Proton therapy uses protons rather than X-rays to treat cancer. Protons can be controlled to stop at a specific depth, thereby minimizing damage to surrounding healthy tissues. In clinical practice, the following ion therapies are currently being implemented:

- *Proton therapy*: It is the most widespread form of ion therapy, with over 100 active centers globally. Its precision in targeting tumors while sparing surrounding healthy tissue makes it a popular choice for various cancer types.

- *Carbon ion therapy*: Approximately, 13 active carbon ion therapy centers exist worldwide. Carbon ions offer higher biological effectiveness than protons, making them suitable for treating radio-resistant tumors.
- *Other ions*: Therapies using ions such as helium or oxygen are less common and primarily available in research settings. These treatments are still being investigated to determine their clinical efficacy and potential advantages.

Table 7.5 lists the typical absorbed doses for ion therapy. Gy(RBE) refers to the absorbed dose of ionizing radiation, measured in Gy, adjusted for its relative biological effectiveness (RBE). This adjustment accounts for the increased biological effectiveness of ion radiation compared to conventional photon-based RT. Typical values for RBE adjustment coefficients used in clinical practice are as follows:

- *Protons'* typical RBE coefficient is approximately 1.1. This reflects their slightly higher biological effectiveness compared to photons, and it serves as a standard value in clinical proton therapy.
- *Carbon ions* have an RBE coefficient of 2.5–3.5. Their high LET makes them particularly effective for treating hypoxic or radio-resistant tumors.
- *Helium ions* have an RBE coefficient of approximately 1.5–2.0. Their LET is higher than that of protons but lower than that of carbon ions, balancing biological effectiveness and precision.
- *Oxygen ions*, which are still primarily experimental, have an RBE coefficient of 3.0–4.0. Their high LET makes them promising for treating radio-resistant tumors, although more research is needed to standardize their use.

Table 7.5. Particle therapy treatments and typical absorbed doses.

Machine	Treatment Approach	Typical Dose Range (Gy)
Ion Therapy	Conventional	1.8–2.0 Gy(RBE) per fraction
	Hypofractionated	2.5–4 Gy(RBE) per fraction
	Stereotactic	5–10 Gy(RBE) per fraction

7.3.1.3 *Intraoperative radiation therapy equipment*

IORT involves delivering a concentrated dose of radiation directly to the tumor bed during surgery. This approach allows for precise tumor targeting while minimizing radiation exposure to surrounding healthy tissues. Several specialized devices are used for IORT, each offering unique treatment delivery and efficiency benefits.

The Mobetron is a portable electron linear accelerator designed for IORT. It delivers high-energy electrons directly to the tumor site during surgery, allowing for precise radiation treatment while the surgical team is present. The LIAC system is a mobile electron linear accelerator for IORT. It delivers high-energy electrons to the target area, focusing on minimizing radiation exposure to nearby healthy tissues.

The Intrabeam System utilizes low-energy X-rays delivered directly to the tumor bed via a miniature X-ray source. It is beneficial for treating early-stage breast cancer during lumpectomy procedures. It represents another option for delivering IORT, particularly for breast cancer treatment. It uses low-energy X-rays from a miniature X-ray source placed directly into the tumor cavity.

7.3.1.4 *Roentgen therapy equipment*

Roentgen therapy, or orthovoltage X-ray therapy, uses X-rays in the range of 50–300 kV to treat superficial tumors and skin cancers. Although it is less commonly used today due to the development of more advanced radiotherapy techniques, it still serves specific purposes, such as treating benign conditions and certain skin cancers.

Tables 7.6 and 7.7 list typical absorbed dose in Gy for IOeRT, Intrabeam, and Roentgen theraphies.

7.3.2 *Conventional delivery strategies*

Modern photon LINACs can deliver diverse delivery techniques, such as 3D conformal radiation therapy (3D-CRT), IMRT, or volumetric

Table 7.6. Particle therapy treatments and typical absorbed doses.

Treatment Approach	Typical Dose Range (Gy)
IOeRT	Typically, higher doses per fraction, ranging from 10 to 25 Gy.
Intrabeam	Single dose ranging from 20 to 50 Gy.

Table 7.7. Roentgen therapy treatments and typical absorbed doses.

Roentgen Therapy	Typical Dose Range (Gy)
Conventional Treatment	1.8–2.5 Gy per fraction, total doses from 0 to 60 Gy over several weeks.
Hypofractionated Treatment	4–6 Gy per fraction, total doses around 20–30 Gy.

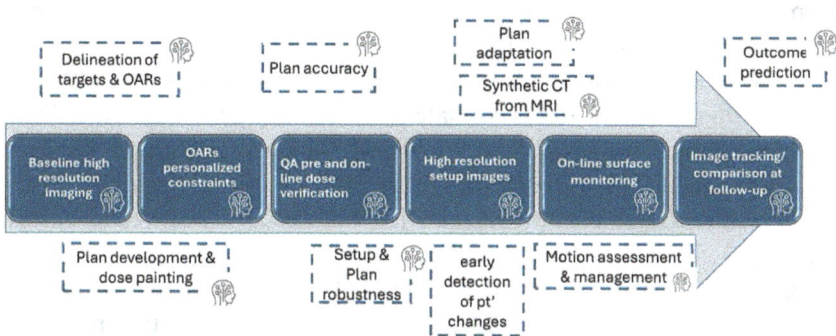

Fig. 7.2. AI-based radiotherapy workflow.

Source: Generated with licensed MS PowerPoint by the authors.

modulated arc therapy (VMAT) to optimize dose distribution, minimize damage to surrounding healthy tissues, and achieve superior treatment outcomes.

Non-standard LINAC delivery techniques are covered in the following sections with illustrative examples of their applications in the radiotherapy workflow (Fig. 7.2).

7.3.2.1 *3D-CRT*

3D-CRT is one of the earliest techniques that integrates imaging data from CT to plan radiation beams, conforming to the tumor's shape [32, 33].

The dose is delivered through multiple static beams, ensuring that the high-dose region closely matches the target volume. Over time, the method evolved by incorporating tools such as wedges and "field-in-field" techniques to improve dose uniformity and reduce hotspots. Although more advanced methods have largely superseded it, 3D-CRT remains a valuable option for certain cancers where precise dose shaping is less critical.

The procedure begins with CT imaging of the patient to identify the tumor and surrounding critical structures. The patient is positioned in a reproducible manner using immobilization devices. Radiation oncologists then outline the target tumor volume, including the gross tumor volume (GTV), clinical target volume (CTV), and planning target volume (PTV), as well as nearby organs at risk (OARs), on the CT images. Physicists use treatment planning systems to position and shape multiple radiation beams around the tumor, selecting beam angles to avoid OARs while delivering a conformal dose [32–34].

Multi-leaf collimators (MLCs) or custom-fitted blocks are employed to sculpt the radiation fields to match the tumor's geometry. Additional techniques, such as wedges or "field-in-field," are applied to improve dose homogeneity. The radiation is delivered in multiple sessions (fractions), with patient positioning verified before each session to ensure accuracy.

One of the main advantages of 3D-CRT is its ability to conform the radiation dose to the tumor shape, significantly reducing exposure to surrounding healthy tissue when compared to earlier 2D methods. The technique is widely available in RT centers. It only requires highly specialized equipment for more advanced techniques such as IMRT [35, 36] or VMAT [37]. It is also cost-effective, making it an accessible option for many patients while providing acceptable treatment outcomes for certain cancers, particularly in countries with limited economic resources. Furthermore, 3D-CRT offers a better toxicity profile than 2D radiotherapy due to its improved dose distribution.

However, 3D-CRT has some notable limitations. While it improves dose conformity over 2D methods, it is less precise than IMRT or stereotactic body radiation therapy (SBRT). Critical organs near the tumor may still receive a higher dose than that achievable with more advanced techniques, thereby increasing the risk of side effects. Additionally, 3D-CRT delivers uniform beam intensity; however, it lacks the ability to vary dose intensity within a beam, which limits its adaptability to complex tumor geometries. Tumors with irregular shapes or those located near sensitive structures may not benefit fully from 3D-CRT due to its less advanced conformality.

Despite its limitations, 3D-CRT remains a valuable and reliable technique in the radiation oncology toolkit, particularly in cases where advanced methods are not accessible or not necessary.

7.3.2.2 *IMRT*

IMRT represents a significant improvement over 3D-CRT because it modulates the intensity of individual radiation beams [34, 35].

This approach allows for highly conformal dose distributions and sparing of adjacent critical structures. IMRT begins with CT or other imaging modalities to delineate the tumor and surrounding normal tissues. Radiation oncologists and physicists identify the target volumes and organs at risk, similar to 3D-CRT. However, IMRT employs advanced software to divide each radiation beam into many small beamlets, each capable of delivering a different dose intensity. These beamlets are optimized using inverse planning algorithms to achieve the desired dose distribution.

The highly conformal nature of IMRT is particularly advantageous for treating tumors with irregular shapes or those located near critical structures, such as the spinal cord or optic nerves. The ability to modulate dose intensity across a single beam allows IMRT to deliver higher doses to the tumor while minimizing exposure to surrounding healthy tissues. This leads to improved tumor control and reduced toxicity. Moreover, IMRT can include simultaneous integrated boost (SIB) techniques, allowing for differential dosing within the target region and providing an additional layer of treatment customization.

Despite its advantages, IMRT has limitations. It is a more time-intensive process than 3D-CRT in treatment planning and delivery. The technique's complexity requires more sophisticated equipment and expertise, which can increase costs and limit availability in some settings. Additionally, the greater number of beam angles and modulated dose delivery may result in a low-dose "bath" for a larger volume of normal tissue, potentially increasing the risk of secondary malignancies over the long term.

IMRT has become a cornerstone in modern radiation oncology, particularly for challenging cases where precision is paramount.

7.3.2.3 *VMAT*

VMAT delivers radiation through continuous arc motion around the patient, dynamically varying the beam's intensity and shape [37–39]. This approach represents an evolution of IMRT, where, instead of static beams, the radiation is delivered as the gantry rotates around the patient. VMAT allows for the use of single or multiple arcs depending on the complexity of the tumor's shape and its proximity to critical structures. Single-arc VMAT is typically sufficient for simpler tumor geometries, while multiple arcs can provide even greater dose conformity and better sparing of adjacent OARs.

The continuous rotation of the gantry during VMAT enables the radiation dose to be distributed more uniformly and efficiently. Advanced planning algorithms control the MLCs, the dose rate, and the gantry speed simultaneously, optimizing radiation delivery to conform to the tumor's shape while minimizing exposure to surrounding healthy tissues. This dynamic modulation allows for steep dose gradients, effectively reducing the dose to nearby OARs and improving the therapeutic ratio.

Due to its efficient delivery mechanism, VMAT offers shorter treatment times than IMRT. This reduction in treatment time improves patient comfort and reduces the risk of patient movement during therapy, which can enhance the overall precision of treatment. However, VMAT requires sophisticated treatment planning systems, high-quality imaging to ensure accuracy, and advanced quality assurance protocols to verify complex delivery parameters.

VMAT is particularly beneficial for treating tumors in anatomically challenging locations, such as head and neck cancers, prostate cancers, and pelvic malignancies, where precision and sparing of OARs are critical [40, 41]. The technique has demonstrated significant advantages in reducing both acute and late toxicities while maintaining excellent tumor control rates.

7.3.2.4 *Non-standard LINACs*

Mackie *et al.* [42] introduced the concept of helical tomotherapy in the early 1990s, based on their work from the late 1980s.

Tomotherapy integrates CT imaging and helical radiation delivery into a single system, combining MV imaging and treatment. This technology uses a fan-shaped radiation beam that continually rotates around the patient, delivering the dose slice by slice. The helical delivery approach allows for precise dose sculpting, making it particularly advantageous for treating complex, irregularly shaped tumors and those located near critical structures. In tomotherapy, the treatment field length in the longitudinal direction can be as large as 40 cm, eliminating the need for field junctions in many cases. This ensures a uniform dose distribution across large target volumes and avoids potential hotspots or underdosed regions at junctions.

Tomotherapy enables multi-isocentric setups for highly complex cases, targeting different tumor regions or multiple tumors with individual precision. Its integrated imaging capabilities, using megavoltage computed tomography (MVCT), provide real-time guidance and ensure accurate patient positioning before and during treatment [43, 44].

Conversely, CyberKnife is a robotic radiosurgery system designed to deliver exact non-coplanar radiation beams [44, 45]. Its unique robotic arm can move with six degrees of freedom, enabling the delivery of beams from virtually any angle. This capability allows CyberKnife to treat tumors with sub-millimeter accuracy, making it ideal for targeting small, moving tumors, such as those in the lungs and livers. The system incorporates real-time image guidance, tracks tumor movement caused by respiration or other physiological processes, and adjusts the beam delivery in real time to maintain accuracy. Unlike traditional linear accelerators, CyberKnife does not require fixed isocenters, allowing greater flexibility in targeting complex geometries [46].

The CyberKnife system can deliver radiation in a frameless setup, making it less invasive for patients than traditional stereotactic radiosurgery methods. It is particularly beneficial for treating lesions in sensitive or hard-to-reach locations, such as brain metastases, spinal tumors, and prostate cancer. The combination of real-time tracking and non-coplanar beam delivery minimizes the dose to surrounding healthy tissue, thereby reducing side effects while achieving high tumor control rates. CyberKnife can be provided with a fixed cone, iris, and MLC [47, 48].

7.3.2.5 *Particle therapy delivery*

Particle therapy is an advanced RT that uses charged particles such as protons, carbon ions, and occasionally neutrons to target cancer with exceptional precision [49].

Unlike conventional photon-based radiotherapy, charged particles deposit their energy in a highly localized manner due to the Bragg peak phenomenon, which enables maximal energy delivery at the tumor site while sparing surrounding healthy tissues.

Proton therapy is widely used for tumors located near critical structures, such as the brain, spinal cord, or optic nerves, and in pediatric cases, where long-term side effects must be minimized. Proton energies typically range from 70 to 250 MeV, allowing them to penetrate deeply while precisely stopping at the desired depth.

Carbon ion therapy is particularly effective for treating radioresistant tumors, such as sarcomas and certain head-and-neck cancers. These tumors are less responsive to conventional radiation but are vulnerable to carbon ions due to their higher LET. This characteristic causes more significant biological damage to cancer cells, enhancing tumor control while reducing damage to healthy tissues.

Though less common, neutron therapy is employed in specific cases, such as salivary gland cancers. Neutrons also have a high LET, which can be advantageous in treating certain tumors. However, their use remains limited due to challenges in dose delivery and higher toxicity to surrounding tissues. Dose delivery in particle therapy relies on advanced technologies to ensure precise treatment. Passive scattering methods broaden the particle beam, shaping it to match the tumor's contour. In contrast, pencil beam scanning (PBS) delivers the beam in a highly focused, layer-by-layer manner, "painting" the tumor with exceptional accuracy. This technique minimizes exposure to adjacent healthy tissues and is especially useful for irregularly shaped or moving tumors [50]. Dose rates in conventional particle therapy typically range from 1 to 10 Gy per minute, allowing for efficient and controlled treatment.

Fractionation schedules in particle therapy vary depending on the particle type and clinical objectives. Proton therapy often uses standard fractionation protocols, delivering 1.8–2 Gy per session. Carbon ion therapy, on the other hand, frequently employs hypofractionated regimens, delivering higher doses per fraction to exploit its superior biological effectiveness. This approach reduces the overall treatment duration while maintaining or improving tumor control.

Particle therapy is particularly effective for cancers near sensitive structures, pediatric malignancies, and radioresistant tumors. Its ability to precisely target tumors while minimizing damage to healthy tissues makes it a preferred option for complex cases. It might reduce the risk of long-term side effects in pediatric patients, such as developmental delays and secondary malignancies.

Recent advances in imaging and beam delivery systems have significantly enhanced the precision and adaptability of particle therapy [51]. Real-time tumor tracking technologies allow for the continuous monitoring of tumor movement caused by respiration or other factors, ensuring accurate dose delivery even to moving targets. Rotating gantries, which deliver beams from multiple angles, further improve treatment flexibility.

Emerging techniques such as FLASH therapy are under active investigation [52]. Delivered at ultrahigh dose rates exceeding 40 Gy/s, FLASH therapy can more effectively spare healthy tissues while maintaining tumor control. This novel approach could revolutionize radiotherapy by reducing treatment toxicity and improving patient outcomes.

Another major innovation is integrating adaptive radiation therapy (ART) into particle therapy. ART combines real-time imaging with treatment adaptation and advanced planning algorithms, allowing for adjustments based on anatomical or physiological changes during treatment. This adaptability ensures optimal dose distribution throughout the therapy. Additionally, arc delivery techniques using charged particle beams, such as protons and carbon ions, are gaining attention for their potential clinical benefits. Delivering the beam in a continuous arc around the patient improves dose conformity to the tumor while reducing exposure to healthy tissues.

As particle therapy continues to evolve, further advancements in imaging, beam delivery, and treatment planning are expected to enhance its clinical utility. Research into novel radiobiological effects, personalized fractionation schedules, and integration with other therapies, such as immunotherapy, is also underway. Developing more cost-effective systems could expand access to this cutting-edge treatment, making it a cornerstone of modern oncology care [53].

7.3.2.6 *Intraoperative*

IORT, introduced in Section 7.3.1.3, is an advanced cancer treatment that delivers high-dose radiation directly to the tumor bed during surgery [54]. This approach reduces the risk of local recurrence while minimizing radiation exposure to surrounding healthy tissues. Depending on the type of radiation, IORT is classified into photon-based and electron-based methods. Each has distinct characteristics regarding machines, energy levels, and dose rates.

Photon-based IORT uses low-energy X-rays produced by compact, mobile units such as the Intrabeam system. These devices typically operate at energies between 50 and 100 kV, offering a shallow penetration, which is ideal for treating superficial tumors or performing partial breast irradiation. Due to their portability, these systems are especially suited for operating rooms and small treatment sites. Because of its lower dose rate, photon-based IORT typically requires longer delivery times than electron-based methods. Still, it remains effective in specific scenarios, especially where flexibility and precision are crucial.

Electron-based IORT (IOeRT) employs LINACs to generate high-energy electron beams. These systems deliver therapeutic energies ranging from 4 to 20 MeV, allowing for a more customized treatment depth based on tumor size and location. LINAC-based systems, such as those produced by Mobetron or LIAC, are larger and often require a shielded operating room. Electrons provide better dose uniformity and can treat more deeply seated tumors compared to photons. Dose rates for IOeRT typically range between 10 and 25 Gy, which can be administered within a few minutes due to the high output of LINACs.

The development of FLASH radiotherapy, particularly with electron beams, has further revolutionized IORT. FLASH-RT involves ultrahigh dose rates exceeding 40 Gy per second, delivered in milliseconds. Preclinical studies suggest that FLASH-RT minimizes damage to healthy tissues while maintaining tumor control, presenting a promising future direction for IORT. These breakthroughs rely on advanced LINACs capable of producing the ultrahigh dose rates necessary for FLASH therapy.

7.3.3 *Innovative delivery strategies*

7.3.3.1 *Lattice and GRID*

Lattice and GRID RT are innovative approaches in radiotherapy aimed at enhancing tumor control while minimizing toxicity to healthy tissues. These techniques differ from conventional radiation delivery methods by using spatially fractionated doses, where radiation is delivered to distinct regions within the tumor rather than uniformly across its entirety. This creates a pattern of high-dose areas (peaks) interspersed with low-dose regions (valleys), enabling selective tumor targeting while sparing normal tissue.

- *Lattice* RT involves delivering high doses of radiation to specific points within the tumor while exposing the surrounding regions to much lower doses. This approach exploits the biological differences between cancer cells and normal tissues. Tumor cells in the high-dose regions experience severe damage, whereas cells in the lower-dose areas still receive enough radiation to contribute to overall tumor control. This selective approach has shown promise in inducing systemic immune responses, potentially enhancing the efficacy of radiation.

- *GRID* RT uses a physical collimator or an advanced multi-leaf collimator to create a grid-like pattern of high-dose regions across the tumor. GRID therapy is often applied with other treatments such as chemotherapy or surgery and is especially useful for treating large tumors. Its ability to deliver high doses without exceeding normal tissue tolerance thresholds has made it an attractive option for challenging cases, including bulky, radioresistant cancers.

7.3.3.2 *FLASH*

FLASH RT, mentioned in the previous paragraphs, is a cutting-edge technique that delivers ultrahigh radiation dose rates, exceeding 40 Gy per second, in a single or a few fractions. With continued research into FLASH therapy and biologically optimized treatments, particle therapy is expected to play an increasingly pivotal role in cancer treatment, offering improved outcomes and reduced side effects.

This method differs significantly from conventional RT, where dose rates are typically much lower, at about 1–10 Gy per minute. The rapid delivery of radiation in FLASH therapy is believed to induce a phenomenon known as the "FLASH effect," which spares healthy tissues while maintaining the effectiveness of tumor control.

The biological basis of the FLASH effect lies in its impact on normal and cancerous tissues. Preclinical studies demonstrate that FLASH irradiation causes less oxidative stress in normal cells than conventional radiation. The extremely short treatment times, often lasting milliseconds, limit the generation of reactive oxygen species (ROS), reducing damage to healthy cells. However, due to their impaired ability to repair DNA damage and higher metabolic demands, cancer cells are less protected by the FLASH effect, leading to effective tumor control.

Regarding technical delivery, FLASH therapy requires highly specialized equipment capable of generating ultrahigh dose rates. Proton accelerators, electron beams, and LINACs modified for FLASH capabilities are commonly used. Electrons have been most extensively studied, though research into FLASH proton and carbon ion therapy is advancing rapidly. Modifying existing linear accelerators for ultrahigh dose rates involves ensuring precise beam control, dose measurement, and real-time monitoring to prevent overexposure.

FLASH therapy's unique dose distribution offers significant advantages in clinical applications. It holds particular promise for treating tumors near critical structures, such as the brain or lungs, where minimizing collateral damage is crucial. Pediatric oncology may also benefit from FLASH therapy due to the reduced long-term toxicity risks of conventional RT. Furthermore, the use of ultrahigh dose rates has the potential to make treatments shorter and more tolerable for patients. Clinical trials are investigating FLASH therapy's safety and efficacy in humans. Early results suggest improved tolerance of normal tissues without compromising tumor control. However, the technique faces challenges, including the need for standardized protocols, dosimetry accuracy, and understanding the mechanisms behind the FLASH effect. Transitioning from preclinical to clinical settings will require rigorous testing to establish optimal parameters for different cancer types and patient populations.

FLASH RT represents a paradigm shift in oncology, merging advances in radiation physics with a deep understanding of tumor biology. If its potential is fully realized, FLASH could significantly improve cancer treatment by reducing toxicity, shortening treatment times, and expanding the therapeutic index of radiotherapy.

7.3.3.3 4π

4π RT is an advanced radiation delivery technique designed to maximize dose conformity to the tumor while sparing surrounding healthy tissues. Named after the mathematical representation of an entire sphere in steradians, 4π radiotherapy leverages the full spatial capabilities of modern LINACs to deliver radiation from nearly every angle around the patient. This results in highly conformal dose distributions and a significant reduction in the dose delivered to adjacent critical structures.

The method uses sophisticated algorithms and real-time imaging to optimize the radiation beams' trajectories. Unlike conventional radiotherapy, which often employs a limited number of fixed angles, 4π radiotherapy incorporates hundreds of non-coplanar beams. This approach ensures that the radiation dose conforms closely to the tumor's shape while avoiding sensitive organs, such as the brainstem, spinal cord, or heart.

4π radiotherapy combined with VMAT enhances dose precision by delivering radiation while modifying the angle of the couch. VMAT's continuous rotation of the LINAC at different couch angles allows dynamic modulation of beam intensity, shape, and dose rate, thereby optimizing the 4π approach. The key to the efficiency of 4π radiotherapy lies in the fact that the 4π algorithms analyze thousands of potential beam angles or arc/couch combinations and select the optimal set to achieve the desired dose distribution.

One of the main applications of 4π radiotherapy is in treating complex tumors near critical structures, such as brain, head and neck, or prostate cancers. By reducing the dose to surrounding healthy tissues, 4π allows for the safe delivery of higher radiation doses to the tumor, potentially improving treatment outcomes. Additionally, it offers benefits in re-irradiation cases, where previous treatments have already exposed healthy tissues to significant radiation.

Despite its advantages, 4π radiotherapy requires advanced equipment and significant computational resources. Treatment planning is time-intensive due to the complex optimization process, and delivery often necessitates highly precise robotic systems. Additionally, patient motion during treatment can pose challenges, requiring advanced immobilization techniques and motion tracking to maintain accuracy.

4π radiotherapy represents a significant advancement in precision oncology. By fully utilizing the spatial flexibility of modern radiotherapy systems, it achieves unparalleled dose conformity, enabling the effective treatment of challenging cases with minimal toxicity. As technology evolves, 4π is expected to become more accessible, offering improved outcomes for a broader range of cancer patients.

7.4 Radiation Therapy Applications Based on Radionuclides

7.4.1 *Sealed radionuclides (brachytherapy)*

Brachytherapy involves placing sealed radioactive sources directly inside or near the tumor. It is highly effective for treating localized cancers, particularly in the prostate, cervix, breast, and skin. It is possible to distinguish three different delivery approaches concerning the used sources, as in the following:

- *Low-dose-rate (LDR)* brachytherapy uses sources that emit radiation at a low rate, typically delivering doses over several days. Common

Table 7.8.   Brachytherapy treatments and typical absorbed doses.

Brachytherapy Approach	Delivery Type	Typical Dose Range (Gy)
Low-Dose-Rate (LDR)	Protracted with permanent seed implants	Cervical cancer: 30–40 Gy combined with EBRT Prostate cancer: 140–160 Gy total
High-Dose-Rate (HDR)	Fractionated	Cervical cancer: 6–7 Gy per fraction, with total doses of 28-30 Gy combined with EBRT Prostate cancer: 8–10 Gy per fraction, usually in 2–4 fractions
Pulsed-Dose-Rate (PDR)	Similar to HDR	Cervical cancer: typically 30–40 Gy in pulsed fractions

isotopes for prostate cancer include iodine-125, cesium-131, and iridium-192 for other applications.

- *High-dose-rate (HDR)* brachytherapy uses sources that deliver radiation at a high rate over a short period. Common isotopes include iridium-192 and cobalt-60.
- *Pulsed-dose-rate (PDR)* brachytherapy combines the advantages of LDR and HDR by delivering radiation in periodic pulses, typically every hour. This technique uses isotopes similar to those used in HDR brachytherapy.

Table 7.8 summarizes typical Brachytheraphy treatments approaches and and correlated patient absorbed dose in Gy.

7.4.2 *Nuclear medicine*

Nuclear medicine therapy, also known as radionuclide therapy, utilizes radioactive substances to treat various medical conditions, including cancer and certain non-malignant diseases. The absorbed dose from these radionuclide-labeling radiopharmaceuticals or those coupled with medical devices shows dose–effect relationships [55], which could guide the treatment personalization. This therapeutic approach involves administering radiopharmaceuticals that selectively target diseased cells, thereby delivering radiation directly to the affected tissues. The concept of theragnostics, which combines therapy and diagnostics, plays a crucial role in nuclear medicine therapy by allowing for personalized treatment strategies based on individual patient characteristics and disease profiles.

In nuclear medicine therapy, radiopharmaceuticals selectively deliver radiation to diseased tissues while minimizing exposure to healthy organs. These radiopharmaceuticals emit various types of radiation, such as beta particles or alpha particles, which can effectively kill or inhibit the growth of target cells. The choice of radiopharmaceutical and administration route depends on factors such as the type and location of the disease, the patient's overall health, and the desired therapeutic outcome.

7.4.2.1 *Theragnostic*

The concept of theragnostics, a combination of "therapy" and "diagnostics," involves using the same radioactive agent for both therapy and diagnosis [56, 57]. This approach identifies suitable patients for therapy based on diagnostic imaging and enables monitoring of treatment response over time. In nuclear medicine therapy, theragnostic pairs consist of diagnostic and therapeutic radiopharmaceuticals targeting the same molecular pathways or receptors. For example, in treating thyroid cancer, diagnostic imaging using Tc-99m pertechnetate or iodine-123 can help localize thyroid tissue and assess thyroid function. This diagnostic information can then inform the decision to administer iodine-131 therapy, which selectively targets thyroid cells and delivers beta radiation to destroy cancerous tissue.

Similarly, PET imaging using gallium-68-labeled somatostatin analogs for treating neuroendocrine tumors can accurately localize tumors expressing somatostatin receptors. This diagnostic information guides the administration of lutetium-177 or actinium-225 therapy, which selectively delivers beta or alpha radiation to tumor cells expressing somatostatin receptors. Theragnostic approaches offer several advantages in nuclear medicine therapy. They allow personalized treatment strategies based on individual patient characteristics and disease profiles. By combining therapy and diagnostics, theragnostic approaches enable precise localization of diseased tissue and assessment of treatment response, leading to optimized therapeutic interventions. Additionally, targeted delivery of radiation to diseased tissue minimizes radiation exposure to healthy organs, thereby reducing the risk of adverse effects.

In conclusion, nuclear medicine therapy, guided by the principles of theragnostics, offers a targeted and personalized approach to treating various medical conditions, including cancer. By combining therapy and diagnostics, theragnostic approaches enable precise localization of diseased tissue, optimization of treatment strategies, and minimization of adverse effects, ultimately improving patient outcomes [58].

References

[1] Tsapaki, V. (2020). Radiation dose optimization in diagnostic and interventional radiology: Current issues and future perspectives. *Physics in Medicine and Biology*, 79, 16–21. https://doi.org/10.1016/j.ejmp.2020.09.015.

[2] Hill, K. D., and Einstein, A. J. (2016). New approaches to reduce radiation exposure. *Trends in Cardiovascular Medicine*, 26(1), 55–65. https://doi.org/10.1016/j.tcm.2015.04.005.

[3] Wagner, M., Schafer, S., Strother, C., and Mistretta, C. (2016). 4D interventional device reconstruction from biplane fluoroscopy. *Medical Physics*, 43(3), 1324–1334. https://doi.org/10.1118/1.4941950.

[4] Bang, J. Y., Hough, M., Hawes, R. H., and Varadarajulu, S. (2020). Use of artificial intelligence to reduce radiation exposure at fluoroscopy-guided endoscopic procedures. *The American Journal of Gastroenterology*, 115(4), 555–561. https://doi.org/10.14309/ajg.0000000000000565.

[5] Cina, A., Steri, L., Barbieri, P., Contegiacomo, A., Amodeo, E. M., Di Stasi, C., Morasca, A., Romualdi, D., Ciccarone, F., and Manfredi, R. (2022). Optimizing the angiography protocol to reduce radiation dose in uterine artery embolization: The impact of digital subtraction angiographies on radiation exposure. *Cardiovascular and Interventional Radiology*, 45(2), 249–254. https://doi.org/10.1007/s00270-021-03032-8.

[6] Nicosia, L., Gnocchi, G., Gorini, I., Venturini, M., Fontana, F., Pesapane, F., Abiuso, I., Bozzini, A. C., Pizzamiglio, M., Latronico, A., Abbate, F., Meneghetti, L., Battaglia, O., Pellegrino, G., and Cassano, E. (2023). History of mammography: Analysis of breast imaging diagnostic achievements over the last century. *Healthcare*, 11(11), 1596. https://doi.org/10.3390/healthcare11111596.

[7] Mileto, A., Yu, L., Revels, J. W., Kamel, S., Shehata, M. A., Ibarra-Rovira, J. J., Wong, V. K., Roman-Colon, A. M., Lee, J. M., Elsayes, K. M., and Jensen, C. T. (2024). State-of-the-art deep learning CT reconstruction algorithms in abdominal imaging. *Radiographics*, 44(12), e240095. https://doi.org/10.1148/rg.240095.

[8] Barca, P., Domenichelli, S., Golfieri, R., Pierotti, L., Spagnoli, L., Tomasi, S., and Strigari, L. (2023). Image quality evaluation of the Precise Image CT deep learning reconstruction algorithm compared to filtered back-projection and iDose4: A phantom study at different dose levels. *Physics in Medicine and Biology*, 106, 102517. https://doi.org/10.1016/j. ejmp.2022.102517.

[9] Tomasi, S., Szilagyi, K. E., Barca, P., Bisello, F., Spagnoli, L., Domenichelli, S., and Strigari, L. (2024). A CT deep learning reconstruction algorithm: Image quality evaluation for brain protocol at decreasing dose indexes compared to FBP and statistical iterative reconstruction algorithms. *Physics in Medicine and Biology*, 119, 103319. https://doi. org/10.1016/j.ejmp.2024.103319.

[10] Greffier, J., Villani, N., Defez, D., Dabli, D., and Si-Mohamed, S. (2023). Spectral CT imaging: Technical principles of dual-energy CT and multi-energy photon-counting CT. *Diagnostic and Interventional Imaging*, 104(4), 167–177. https://doi.org/10.1016/j.diii.2022.11.003.

[11] Gullberg, G. T., Christian, P. E., and Zeng, G. S. L. (1991). Cone beam tomography of the heart using single-photon emission-computed tomography. *Investigative Radiology*, 26, 681–688. https://doi.org/10.1097/ 00004424-199107000-00014.

[12] Lalush, D. S., and Tsui, B. M. W. (1998). Block-iterative techniques for fast 4D reconstruction using a priori motion models in gated cardiac SPECT. *Physics in Medicine and Biology*, 43, 875–886. https://doi.org/10.1088/ 0031-9155/43/4/015.

[13] Chen, J., Garcia, E. V., Folks, R. D., *et al.* (2005). Onset of left ventricular mechanical contraction as determined by phase analysis of ECG-gated myocardial perfusion SPECT imaging: Development of a diagnostic tool for assessment of cardiac mechanical dyssynchrony. *Journal of Nuclear Cardiology*, 12, 687–695. https://doi.org/10.1016/j.nuclcard.2005.06.088.

[14] Garcia, E. V., Faber, T. L., and Esteves, F. P. (2011). Cardiac dedicated ultrafast SPECT cameras: New designs and clinical implications. *Journal of Nuclear Medicine*, 52, 210–217. https://doi.org/10.2967/jnumed. 110.081323.

[15] Nicolas, J. M., Catafau, A. M., Estruch, R., *et al.* (1993). Regional cerebral blood flow-SPECT in chronic alcoholism – Relation to neuropsychological testing. *Journal of Nuclear Medicine*, 34, 1452–1459.

[16] Sanna, G., Piga, M., Terryberry, J. W., *et al.* (2000). Central nervous system involvement in systemic lupus erythematosus: Cerebral imaging and serological profile in patients with and without overt neuropsychiatric manifestations. *LUPUS*, 9, 573–583. https://doi.org/10.1191/096120300 678828695.

[17] Perri, M., Erba, P., Volterrani, D., *et al.* (2008). Octreo-SPECT/CT imaging for accurate detection and localization of suspected neuroendocrine tumors. *Quarterly Journal of Nuclear Medicine and Molecular Imaging*, 52, 323–333.

[18] Keidar, Z., Israel, O., and Krausz, Y. (2003). SPECT/CT in tumor imaging: Technical aspects and clinical applications. *Seminars in Nuclear Medicine*, 33, 205–218. https://doi.org/10.1053/snuc.2003.127310.

[19] Christian, J. A., Partridge, M., Mioutskkou, E., *et al.* (2005). The incorporation of SPECT functional lung imaging into inverse radiotherapy planning for non-small lung cancer. *Radiotherapy and Oncology*, 77, 271–277. https://doi.org/10.1016/j.radonc.2005.08.008.

[20] Lerman, H., Metser, U., Lievshitz, G., *et al.* (2006). Lymphoscintigraphic sentinel node identification in patients with breast cancer: The role of SPECT-CT. *European Journal of Nuclear Medicine and Molecular Imaging*, 33, 329–337. https://doi.org/10.1007/s00259-005-1927-4.

[21] Bhushan, K. R., Misra, P., Liu, F., *et al.* (2008). Detection of breast cancer microcalcifications using a dual-modality SPECT/NIR fluorescent probe. *Journal of the American Chemical Society*, 130, 17648–17649. https://doi.org/10.1021/ja807099s.

[22] Fukuyama, H., Ouchi, Y., Matsuzaki, S., *et al.* (1997). Brain functional activity during gait in normal subjects: A SPECT study. *Neuroscience Letters*, 228, 183–186. https://doi.org/10.1016/S0304-3940(97)00381-9.

[23] Shao, W., Rowe, S. P., and Du, Y. (2021). Artificial intelligence in single photon emission computed tomography (SPECT) imaging: A narrative review. *Annals of Translational Medicine*, 9(9), 820. https://doi.org/10.21037/atm-20-5988.

[24] Dietze, M. M. A., Branderhorst, W., Kunnen, B., *et al.* (2019). Accelerated SPECT image reconstruction with FBP and an image enhancement convolutional neural network. *EJNMMI Physics*, 6.

[25] Floyd, C. R. (1991). An artificial neural network for SPECT image reconstruction. *IEEE Transactions on Medical Imaging*, 10(4), 485–487. https://doi.org/10.1109/42.97600.

[26] Strigari, L., Marconi, R., and Solfaroli-Camillocci, E. (2023). Evolution of portable sensors for in-vivo dose and time-activity curve monitoring as tools for personalized dosimetry in molecular radiotherapy. *Sensors*, 23(5), 2599. https://doi.org/10.3390/s23052599.

[27] Cook, G. J. R., Alberts, I. L., Wagner, T., Fischer, B. M., Nazir, M. S., and Lilburn, D. (2024). The impact of long axial field of view (LAFOV) PET on oncologic imaging. *European Journal of Radiology*, 183, 111873. https://doi.org/10.1016/j.ejrad.2024.111873.

[28] Pratte, J. F., Nolet, F., Parent, S., Vachon, F., Roy, N., Rossignol, T., Deslandes, K., Dautet, H., Fontaine, R., and Charlebois, S. A. (2021).

3D photon-to-digital converter for radiation instrumentation: Motivation and future works. *Sensors*, 21(2), 598. https://doi.org/10.3390/s21020598.

[29] Jimenez-Mesa, C., Arco, J. E., Martinez-Murcia, F. J., Suckling, J., Ramirez, J., and Gorriz, J. M. (2023). Applications of machine learning and deep learning in SPECT and PET imaging: General overview, challenges, and future prospects. *Pharmacological Research*, 197, 106984. https://doi.org/10.1016/j.phrs.2023.106984.

[30] Shiyam Sundar, L. K., Gutschmayer, S., Maenle, M., and Beyer, T. (2024). Extracting value from total-body PET/CT image data – The emerging role of artificial intelligence. *Cancer Imaging*, 24(1), 51. https://doi.org/10.1186/s40644-024-00684-w.

[31] Li, J., Ye, C., Li, S., and Lin, G. (2024). Mapping the evolution of PET/MR research: A bibliometric analysis of publication trends, leading contributors, and conceptual frameworks (2011–2023). *EJNMMI Research*, 8(1), 36. https://doi.org/10.1186/s41824-024-00224-6.

[32] Bentel, G. C., *et al.* (1999). *Radiation Therapy Planning* (2nd ed.). McGraw-Hill Professional, New York.

[33] Khan, F. M., *et al.* (2014). *The Physics of Radiation Therapy* (5th ed.). Wolters Kluwer Health, Philadelphia.

[34] Webb, S., and Powers, W. (1994). The physics of three-dimensional radiation therapy, conformal radiation therapy, radiosurgery, and treatment planning. *Physics Today*, 47(6), 75–76. https://doi.org/10.1063/1.2808536.

[35] Verhey, L. J. (1999). Comparison of three-dimensional conformal radiation therapy and intensity-modulated radiation therapy systems. *Seminars in Radiation Oncology*, 9(1), 78–98. https://doi.org/10.1016/s1053-4296(99)80056-3.

[36] Bortfeld, T. (2006). IMRT: A review and preview. *Physics in Medicine and Biology*, 51(13), R363–R379. https://doi.org/10.1088/0031-9155/51/13/R21.

[37] Schulz, R. J., and Kagan, A. R. (2002). On the role of intensity-modulated radiation therapy in radiation oncology. *Medical Physics*, 29(7), 1473–1482. https://doi.org/10.1118/1.1487859.

[38] Otto, K. (2008). Volumetric modulated arc therapy: IMRT in a single gantry arc. *Medical Physics*, 35(1), 310–317. https://doi.org/10.1118/1.2818738.

[39] Teoh, M., Clark, C. H., Wood, K., Whitaker, S., and Nisbet, A. (2011). Volumetric modulated arc therapy: A review of current literature and clinical use in practice. *British Journal of Radiology*, 84(1007), 967–996. https://doi.org/10.1259/bjr/22373346.

[40] Zhang, P., Happersett, L., Hunt, M., Jackson, A., Zelefsky, M., and Mageras, G. (2010). Volumetric modulated arc therapy: Planning and evaluation for prostate cancer cases. *International Journal of Radiation Oncology Biology Physics*, 76(5), 1456–1462. https://doi.org/10.1016/j.ijrobp.2009.01.054.

[41] Schulz, R. J., and Kagan, A. R. (2002). On the role of intensity-modulated radiation therapy in radiation oncology. *Medical Physics*, 29(7), 1473–1482. https://doi.org/10.1118/1.1487859.

[42] Mackie, T. R., Holmes, T., Swerdloff, S., Reckwerdt, P., Deasy, J. O., Yang, J., Paliwal, B., and Kinsella, T. (1993). Tomotherapy: A new concept for the delivery of dynamic conformal radiotherapy. *Medical Physics*, 20(6), 1709–1719. https://doi.org/10.1118/1.596958.

[43] Sen, A., and West, M. K. (2009). Commissioning experience and quality assurance of helical tomotherapy machines. *Journal of Medical Physics*, 34(4), 194–199. https://doi.org/10.4103/0971-6203.56078.

[44] Chen, S., Wang, J., Hu, W., and Xu, Y. (2024). Comparative evaluation of dosimetric quality and treatment efficiency for Halcyon, TrueBeam, and TomoTherapy in cervical-thoracic esophageal cancer radiotherapy. *Technology in Cancer Research & Treatment*, 23, 15330338241293321. https://doi.org/10.1177/15330338241293321.

[45] Adler, J. R. Jr., Chang, S. D., Murphy, M. J., Doty, J., Geis, P., and Hancock, S. L. (1997). The Cyberknife: A frameless robotic system for radiosurgery. *Stereotactic and Functional Neurosurgery*, 69(1-4 Pt 2), 124–128. https://doi.org/10.1159/000099863.

[46] Gibbs, I. C. (2006). Frameless image-guided intracranial and extracranial radiosurgery using the Cyberknife robotic system. *Cancer Radiotherapy*, 10(5), 283–287. https://doi.org/10.1016/j.canrad.2006.05.013.

[47] Biasi, G., Petasecca, M., Guatelli, S., Martin, E. A., Grogan, G., Hug, B., Lane, J., Perevertaylo, V., Kron, T., and Rosenfeld, A. B. (2018). CyberKnife® fixed cone and Iris™ defined small radiation fields: Assessment with a high-resolution solid-state detector array. *Journal of Applied Clinical Medical Physics*, 19(5), 547–557. https://doi.org/10.1002/acm2.12414.

[48] Jang, S. Y., Lalonde, R., Ozhasoglu, C., Burton, S., Heron, D., and Huq, M. S. (2016). Dosimetric comparison between cone/Iris-based and InCise MLC-based CyberKnife plans for single and multiple brain metastases. *Journal of Applied Clinical Medical Physics*, 17(5), 184–199. https://doi.org/10.1120/jacmp.v17i5.6260.

[49] Chandra, R. A., Keane, F. K., Voncken, F. E. M., and Thomas, C. R. Jr. (2021). Contemporary radiotherapy: Present and future. *Lancet*, 398(10295), 171–184. https://doi.org/10.1016/S0140-6736(21)00233-6.

[50] Yap, J., De Franco, A., and Sheehy, S. (2021). Future developments in charged particle therapy: Improving beam delivery for efficiency and efficacy. *Frontiers in Oncology*, 11, 780025. https://doi.org/10.3389/fonc.2021.780025.

[51] Knäusl, B., Belotti, G., Bertholet, J., Daartz, J., Flampouri, S., Hoogeman, M., Knopf, A. C., Lin, H., Moerman, A., Paganelli, C.,

Rucinski, A., Schulte, R., Shimizu, S., Stützer, K., Zhang, X., Zhang, Y., and Czerska, K. (2024). A review of the clinical introduction of 4D particle therapy research concepts. *Physics Imaging and Radiation Oncology*, 29, 100535. https://doi.org/10.1016/j.phro.2024.100535.

[52] Atkinson, J., Bezak, E., Le, H., and Kempson, I. (2023). The current status of FLASH particle therapy: A systematic review. *Physics in Engineering and Science Medicine*, 46(2), 529–560. https://doi.org/10.1007/s13246-023-01266-z.

[53] Mein, S., Wuyckens, S., Li, X., Both, S., Carabe, A., Vera, M. C., Engwall, E., Francesco, F., Graeff, C., Gu, W., Hong, L., Inaniwa, T., Janssens, G., de Jong, B., Li, T., Liang, X., Liu, G., Lomax, A., Mackie, T., Mairani, A., Mazal, A., Nesteruk, K. P., Paganetti, H., Pérez Moreno, J. M., Schreuder, N., Soukup, M., Tanaka, S., Tessonnier, T., Volz, L., Zhao, L., and Ding, X. (2024). Particle arc therapy: Status and potential. *Radiotherapy and Oncology*, 199, 110434. https://doi.org/10.1016/j.radonc.2024.110434.

[54] Calvo, F. A., Serrano, J., Cambeiro, M., Aristu, J., Asencio, J. M., Rubio, I., Delgado, J. M., Ferrer, C., Desco, M., and Pascau, J. (2022). Intra-operative electron radiation therapy: An update of the evidence collected in 40 years to search for models for electron-FLASH studies. *Cancers*, 14(15), 3693. https://doi.org/10.3390/cancers14153693.

[55] Strigari, L., Konijnenberg, M., Chiesa, C., Bardies, M., Du, Y., Gleisner, K. S., Lassmann, M., and Flux, G. (2014). The evidence base for the use of internal dosimetry in the clinical practice of molecular radio-therapy. *European Journal of Nuclear Medicine and Molecular Imaging*, 41(10), 1976–1988. https://doi.org/10.1007/s00259-014-2824-5.

[56] Laudicella, R., Albano, D., Annunziata, S., Calabrò, D., Argiroffi, G., Abenavoli, E., Linguanti, F., Albano, D., Vento, A., Bruno, A., Alongi, P., and Bauckneht, M. (2019). Theragnostic use of radiolabelled Dota-peptides in meningioma: From clinical demand to future applications. *Cancers*, 11(10), 1412. https://doi.org/10.3390/cancers11101412.

[57] Nyakale, N. E., Aldous, C., Gutta, A. A., Khuzwayo, X., Harry, L., and Sathekge, M. M. (2023). Emerging theragnostic radionuclide applications for hepatocellular carcinoma. *Frontiers in Nuclear Medicine*, 3, 1210982. https://doi.org/10.3389/fnume.2023.1210982.

[58] Kiess, A. P., O'Donoghue, J., Uribe, C., and Lim, K. (2022). Theragnostic use of radiolabelled Dota-peptides in meningioma: From clinical demand to future applications. *Cancers*, 11(10), 1412. https://doi.org/10.3390/cancers11101412.

Chapter 8

Synergies between Radiobiology of Space and Earth

8.1 Introduction

Building on the foundational knowledge presented in previous chapters, this one will explore the similarities and differences in radiobiology between space and Earth environments and their potential implications for human health. Also, a key topic is creating new research disciplines, beginning with well-established ones. In that sense, starting from an epistemological consideration [1], we focus on the differences between multidisciplinary and interdisciplinary research. Another key point will be to realize whether space radiobiology research should be considered multidisciplinary or interdisciplinary. Finally, we return to the primary purpose that gives the rationale for this book: examining these aspects through a case study of research conducted by the AMS research group at the INFN, Roma I division (Sapienza University)—hereafter referred to as the INFN Roma Sapienza AMS research group Rome. Italy—in collaboration with the Medical Physics Department at the IRCCS University Hospital S. Orsola, Bologna Italy. A key point of this lies in the examination and description of the research paths that will raise their aims and prospects for the future.

8.1.1 *Multidisciplinary research*

Multidisciplinary research involves collaboration among teams of scientists from various disciplines, each contributing to a research project while adhering to their respective frameworks, traditions, and expertise. This approach enables them to leverage diverse knowledge, skills, and technologies familiar to their fields. While maintaining their disciplinary focus, these teams integrate their efforts with those of others, forming a collective strategy to address complex problems.

This method is particularly prominent in fundamental and applied research contexts, especially in large-scale projects where multiple groups work on distinct aspects of a broader goal. Examples include initiatives such as developing the "Starship" at SpaceX and constructing the Large Hadron Collider, a high-energy particle accelerator, at the CERN laboratory organization. In such projects, teams focus on specific design and engineering challenges, guided by prior scientific, technical, and political or business objectives. This pooling of specialized efforts fosters innovative solutions to multifaceted problems.

8.1.2 *Interdisciplinary research*

Interdisciplinary research represents a shift toward integrating methodologies, technologies, and perspectives from multiple disciplines to form new hybrid fields. This approach is driven by the need to tackle complex problems that transcend traditional boundaries and is further enabled by technological advancements that reshape and merge existing fields. However, the process demands navigating challenges associated with disciplinary vocabularies, traditions, and worldviews. These difficulties extend beyond scientific practice; they require changes in the policies and structures of research organizations to support interdisciplinary work effectively.

Unlike multidisciplinary research, which involves distinct teams working on separate aspects of a problem within their established frameworks, interdisciplinary research fosters a synthesis of elements from diverse disciplines. Teams collaborate deeply, learning each other's languages and methodologies to develop novel, impossible approaches within any discipline. This process often involves designing experiments that bridge disciplines, creating datasets that interact across fields, and ensuring a shared understanding among team members to validate these connections.

The organizational challenges of interdisciplinary research include breaking down departmental silos, recognizing team leadership, cross-training researchers, and restructuring administrative practices to promote collaborative inquiry. Such transformations are crucial for fostering an academic culture that prioritizes interdisciplinary approaches. Research consortia addressing health challenges resistant to traditional methods exemplify this model by integrating biomedical, clinical, behavioral, and social sciences. These efforts generate innovative solutions and catalyze cultural and structural changes in research environments.

The field of space radiobiology exemplifies an interdisciplinary research organization at its core, blending expertise across a spectrum of scientific domains such as biology, physics, astrophysics, planetary science, and engineering. Researchers must integrate methodologies and perspectives from these distinct fields to comprehensively understand how ionizing radiation (IR) interacts with biological systems and to devise practical solutions to the challenges posed by the space environment. This collaborative effort is a methodological necessity and a defining feature of success when tackling complex questions within this domain.

Since 2017, significant AMS INFN Roma Sapienza research group efforts have underscored the importance of interdisciplinary frameworks. The activities undertaken will be the objects of the following sections. These efforts highlight the evolution of radiobiological studies from compartmentalized disciplines into a more unified approach, where cross-disciplinary collaboration drives innovation. By bridging gaps between fundamental biological studies, high-energy physics experiments, advanced engineering designs, and the astrophysical characterization of space radiation (SR) environments, this research provides a holistic understanding critical for advancing space exploration and ensuring the safety of astronauts on long-duration missions. This interdisciplinary synergy serves as both a model and a necessity for advancing our understanding of radiation effects in space. It lays the groundwork for scientific breakthroughs that can protect human explorers and expand our reach into the cosmos.

8.1.3 *Moving ahead toward different research*

The decision to venture into space radiobiology research, leveraging the knowledge and data from the AMS-02 cosmic ray detector (CRD), was shaped by extensive discussions with colleagues. These conversations provided the foundation of purpose and motivation for this interdisciplinary journey. Three primary factors influenced the decision.

First, the quality of the research environment provided by the AMS experiment was unparalleled. AMS represents excellence in CR measurement for fundamental physics, supported by sophisticated instruments and a highly skilled collaborative team. This environment reflects the legacy of high-energy nuclear and particle physics, a tradition rooted in the INFN Roma Sapienza division in the groundbreaking work by Enrico Fermi and the "Panisperna boys." A parallel to this tradition is found in medical physics, where day-to-day research and dedication to advancing patient care have been pivotal at institutions. The interdisciplinary connection between fundamental physics and applied medical sciences expertise created a robust foundation for exploring space radiobiology.

Second, the encouragement to "think differently" was a recurring theme in discussions with peers. This principle highlights the necessity of exploring uncharted territories in research, stepping beyond the comfort zone of established norms. In science, every answer prompts a new question, and pursuing such unexplored paths is essential for innovation and discovery.

Finally, the "just-in-time" nature of the topic underscores its relevance [2]. By 2018, the resurgence of plans for crewed deep-space exploration, dormant since the Apollo missions of the 1970s, reinvigorated interest in understanding SR effects. This shift aligned with a renewed focus on research areas that had seen limited progress during the intervening decades. Addressing these challenges became a pressing and timely priority, emphasizing the significance of applied research in human space exploration.

These motivations collectively served as the driving force behind integrating expertise from various disciplines, fostering a novel approach to investigating the biological effects of SR and their implications for future missions.

8.2 Case Study: The INFN Roma Sapienza AMS Research Group

This brief section provides a specific case study, or example, related to SR research conducted by The INFN Roma Sapienza AMS research group.

We also describe the connections between research in these fields and possible synergies.

8.2.1 *Bridging the gap: finding a common (new) language*

As mentioned in Section 8.1.2, establishing a common and innovative language is one of the fundamental initial steps in interdisciplinary research. This objective was also among the earliest collaborative efforts, bringing together researchers from diverse scientific communities and organizations through a series of meetings and collaborative sessions. Since 2018, the INFN Roma Sapienza AMS group has collaborated with researchers and scientists to investigate the possibilities of using the CRD to improve the process of health risk assessment due to ionizing radiation for humans in deep space missions. Collaborations focused on creating synergy within different scientific communities (radiobiology, medical physics, radiotherapy, and nuclear medicine) and institutions (research, universities, and national space agencies).

Then, we conducted many studies [3–5] on the capabilities and possibilities in that direction, especially regarding AMS02. We also identified many opportunities for improvement [6], starting from an analysis of the required steps for human health risk assessment in space missions concerning exposure to SR. The fundamental idea is that for each future space mission, especially those in deep space, it will be necessary to carefully consider the mission profile, including the final destination, travel, and duration of exposure, to derive a prediction method for assessing the risks associated with IR.

8.2.2 *Space mission risk assessment: health risks due to ionizing radiation*

Assessing health risks due to IR during space missions is a multi-step process that integrates environmental modeling, biological data, computational analysis, and mission-specific scenarios. This structured approach ensures the safety and well-being of astronauts exposed to high-energy radiation in space. The following outlines the required steps to evaluate and mitigate these risks.

8.2.2.1 *Data collection*

The process begins with collecting data on SR using instruments and detectors. These detectors, located on spacecraft and ground-based platforms, measure cosmic radiation's energy, types, and intensities.

This information serves as the foundation for building detailed radiation field models, capturing the spectrum and distribution of high-energy particles.

8.2.2.2 *Environmental modeling*

Environmental modeling is the next step, in which collected data from CRDs are used to simulate the radiation environment astronauts will encounter in space. This model incorporates several critical factors. Solar activity, for example, plays a significant role in modulating radiation levels, as the Sun's cycles affect the flux and energy of particles in space. The geomagnetic shielding provided by Earth's magnetic field is also considered, as it deflects many high-energy particles and reduces radiation exposure near our planet.

Another crucial aspect is spacecraft shielding, which examines how radiation interacts with the spacecraft's materials. These interactions often produce secondary radiation, such as neutrons or gamma rays, which can significantly contribute to the overall exposure. By integrating all these variables, the environmental model generates a detailed map of the radiation field, encompassing both primary CRs and secondary radiation. This model serves as the foundation for dose calculations and mission-specific risk assessments.

8.2.2.3 *Dose calculation*

Dose calculation involves determining how radiation interacts with the human body and estimating the absorbed radiation dose for astronauts. This is achieved using transport codes and mathematical models that simulate the movement and interaction of radiation particles as they pass through materials and biological tissues. These codes account for the unique properties of each type of radiation, such as protons, heavy ions, and secondary particles, collectively referred to as radiation quality.

Another critical factor in dose calculation is tissue radiosensitivity. Different tissues in the human body respond differently to radiation, with some being more vulnerable to damage than others. Computational phantoms are employed to model these interactions accurately. These digital anatomical models replicate human anatomy and precisely represent organs and tissues. The absorbed dose for specific organs can be accurately calculated using computational phantoms.

This step translates raw environmental data into a detailed picture of how radiation exposure affects various body parts, providing essential inputs for biological effect modeling and risk assessment.

8.2.2.4 *Biological effect modeling*

Radiation's biological impact is not solely determined by the dose absorbed; the type of radiation also plays a role. This is accounted for using the relative biological effectiveness (RBE), a parameter that adjusts the absorbed dose based on the radiation's ability to cause biological damage. The effective dose is calculated by combining the absorbed dose, radiation quality factors (QFs), and tissue-specific sensitivity data. This approach translates physical measurements into biologically relevant terms, bridging the gap between physics and health outcomes.

8.2.2.5 *Space exposure scenarios*

Every mission involves unique conditions that influence radiation exposure. Key factors include the following:

- *Where:* The mission location (e.g., Earth orbit, lunar surface, or deep space).
- *How long*: The duration of astronaut exposure to the radiation environment.
- *When*: Timing of the mission to solar cycles or periods of heightened CR activity.

These parameters define the exposure scenarios, enabling mission planners to anticipate and model the radiation risks specific to the mission's timeline and trajectory.

8.2.2.6 *Experimental validation*

Ground- and space-based experiments are critical for validating the models and calculations. These include accelerator experiments simulating cosmic radiation interactions in laboratory settings to study biological effects and refine computational models.

Experimental validation ensures that the risk assessment process is rooted in empirical evidence and continuously improved.

8.2.2.7 *Risk characterization*

The next step is to apply dose–effect models to estimate potential health risks. These risks include acute effects and short-term impacts, such as radiation sickness or tissue damage, as well as chronic effects and long-term risks, including cancer and degenerative diseases.

Risk characterization involves quantifying the likelihood and severity of these outcomes, offering a detailed understanding of how radiation exposure affects astronaut health.

8.2.2.8 *Risk mitigation*

Risk mitigation involves developing strategies to reduce the potential health impacts of radiation exposure on astronauts. One key approach is optimizing spacecraft shielding, which requires a careful balance between minimizing radiation exposure and managing weight constraints. Shielding materials and configurations are designed to absorb or deflect harmful radiation while ensuring the spacecraft remains efficient and functional. Operational adjustments also play an essential role in mitigation. For instance, mission planners can schedule spacewalks or other high-risk activities during periods of lower radiation exposure, such as times of reduced solar activity. This timing helps limit astronauts' cumulative dose during the mission. In addition, pharmacological countermeasures are explored as part of the mitigation strategy. These include drugs that may protect cells from radiation damage or promote repair mechanisms in affected tissues.

By combining these approaches, a comprehensive risk management plan is developed. This plan integrates shielding solutions, mission timing, and medical countermeasures to ensure astronaut safety while achieving mission objectives.

Assessing health risks from IR during space missions is a multidisciplinary effort that combines physics, biology, and engineering. Mission planners can mitigate radiation risks by systematically collecting data, modeling radiation environments, calculating doses, and validating findings through experiments. This comprehensive approach ensures that astronauts are protected as they venture into the most extreme environments humanity has ever explored.

The following sections focus on how data collected with CRDs can be used to improve the risk assessment process. Figure 8.1 schematically represents the process and potential improvements brought about by CRDs.

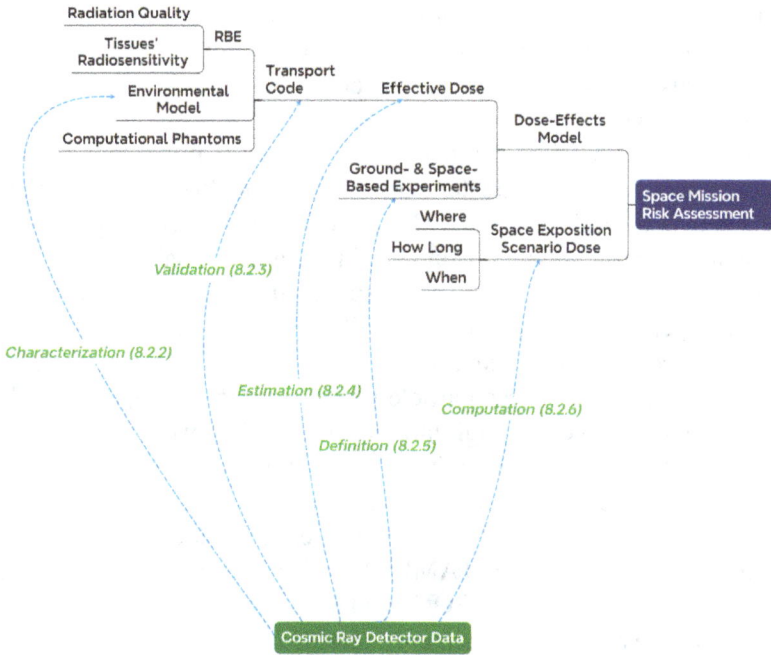

Fig. 8.1. Human health risk assessment in exploratory space missions: possible improvements using astroparticle experiments operating in space.

Source: Generated with a licensed Xmind tool by the authors.

8.2.3 *Environmental model characterization*

The need for an accurate model that describes galactic cosmic rays (GCRs) and the radiation environment is becoming increasingly important in light of planned space exploration missions. The radiation environment in low Earth orbits is the free-space environment, modified by Earth's geomagnetic field. Thus, for interplanetary and low Earth orbital missions, the accuracy of the free-space environment is essential. Attempts to model this environment began in the early 1980s, leading to workable GCR models, called cosmic radiation environmental models (CREMs). Models of the radiation environment in free space and near-Earth orbits are required to estimate the radiation dose to astronauts during Mars, Space Shuttle, and ISS missions and to estimate the rate of single event upsets and latch-ups in electronic devices. Accurate knowledge of the environment is critical for designing optimal shielding during both the cruise phase and for a Mars or Moon habitat. The energy spectra of GCRs have been measured for nearly four decades. In the past decade, models have been constructed that can predict the energy spectrum of any GCR nucleus with greater than 25% accuracy [7].

Current CREMs used in the risk assessment process are based on a subset of the CR spectrum that needs to be improved in information about CR components of energy greater than 1 GeV, owing to limited information collected over the past few years. This affects the accuracy and precision of the risk assessment, potentially underestimating the actual damage.

Many successful CR observatory space missions have collected crucial data over the past decade and will continue doing so. These data exhibit unprecedented precision in the spectral and linear energy transfer (LET) distribution of charge particle (CP) fluxes that compose the CRs. The precision achieved through monitoring the CR fluxes and their variations over time (including the frequency and duration of solar events) is essential for improving risk assessment models.

In this context, Slaba and Witham [8] present a potential improvement by utilizing the AMS-02 and PAMELA CRDs data within the Badhwar–O'Neill (BON) CREM. The BON model has been widely employed to characterize the cosmic radiation encountered in deep space by astronauts and sensitive electronics. The updated BON model incorporates novel methods for computing the solar modulation potential, utilizing new data from AMS-02 and PAMELA to calibrate local interstellar spectrum (LIS) parameters. The previously described uncertainty quantification techniques demonstrate that BON2020 significantly outperforms its predecessor, BON2014. The propagation of uncertainties shows that these updates lead to reduced errors in the effective dose equivalent compared to BON2014.

Another example pertains to the validation of the neutron monitor (NM) network using astroparticle experiment data [9]. The flux of GCRs near Earth is subject to solar modulation, which is influenced by solar magnetic activity and geomagnetic shielding. The global network of ground-based NMs has continuously monitored the variability of GCR flux since the 1950s. Solar modulation is often quantified through the force-field approximation, parameterized by the modulation potential ϕ, which can be evaluated from the comprehensive NM dataset. The methodology has been revised, leading to an updated and extended reconstruction of the solar modulation potential for 1964–2022. This reconstruction incorporates a recent NM yield function and benefits from recent measurements conducted with the AMS-02 [10].

Many other models and tools have been developed in the past decades. Table 8.1 lists the above-mentioned and others not included for

Table 8.1. Comprehensive summary of galactic cosmic ray (GCR) models.

Model Name and Year	Key Features	Applications
Badhwar–O'Neill (BON), Updated 2020	Incorporates AMS-02 and PAMELA data, improved solar modulation potential computation, and reduced uncertainty in effective dose equivalent.	Used for astronaut risk assessment and electronics in deep space, Mars, and Moon mission planning.
CREME (CREME86, CREME96, 1996)	Simulates worst-case and average GCR fluxes, with improved solar modulation modeling in CREME96 – extended energy range.	Widely used for radiation effects analysis on electronics in space.
Nymmik Model, 1996	It considers time-varying solar modulation and includes anomalous GCR components.	Useful for GCR flux predictions near Earth and mission assessments aligned with ISO standards.
ISO-15390, 2004	Defines energy spectra of GCRs at 1 AU with solar modulation parameters.	Serves as a benchmark for GCR spectrum modeling and mission design studies.
Force-Field Model, ~1960s	Simplifies heliospheric modulation using a single potential parameter (ϕ).	Widely employed for near-Earth GCR flux representation and energy distribution analysis.
SPENVIS Tools, Updated 2024	Comprehensive radiation analysis toolkit, integrating models like CREME for simulations in interplanetary and near-Earth environments.	Used for spacecraft radiation environment studies and mission planning.
IRENE Model, Updated 2020	It focuses on trapped particle radiation environments near Earth, complementing GCR models.	Supports spacecraft environment testing and mission design.
Local Interstellar Spectrum (LIS), Modern, 2015	Defines interstellar spectra for calibrating CREMs like BON2020. Tracks cosmic flux variations due to solar modulation.	Improves shielding and dose prediction for interplanetary missions.
Neutron Monitor-Based Solar Modulation, 2022	It uses neutron monitor networks to reconstruct long-term solar modulation potential and tracks GCR flux variability near Earth.	Enables historical and future mission planning by understanding long-term trends in solar modulation.

reference as potential tools or models that may be improved in the future using CRD data.

8.2.4 *Monte Carlo-based transport codes validation*

Monte Carlo (MC) simulation, transport code studies and validations, and cross-section measurements are primarily conducted on the ground, not in space. Based on the detailed information obtained from the CRDs, the MC code can be further refined to better describe interactions with the matter of GCR environments due to improved accuracy of cross-sections at high energies of elementary particles (electrons, protons), light, and heavy nuclei (HN) (from helium to iron and beyond). Implementing transport codes at these energies allows for predicting particle interactions with the known geometries of installed detectors. Determining ray/particle tracking, energy spectra, and deposited energies collected in several materials can serve as a subsequent MC transport code validation (e.g., through a possible Bayesian approach). The calculations of dose equivalents allow for the generation of an accurate and precise database for subsequent MC simulation code validation applied to human tissue. Moreover, refined MC codes can be used to design *ad hoc* shielding for spacecraft and space landers. Considering the importance of radiation dosimetry and the awareness of biological damage caused by the collision of SR on human body tissue, it is necessary to use MC simulation codes for studying particle transport in materials.

For example, the AMS collaboration utilized precise measurements of daily fluxes of GCR electrons and protons from 2011 to 2021 to enhance our understanding of the Sun's influence on CRs [11]. The study revealed distinctive variations in electron fluxes across multiple timescales, which differed significantly from those observed in proton fluxes over the same 11-year period. These findings will enable the investigation of GCR drift effects, particularly the disruption induced by turbulence, at an unprecedented level of accuracy. Furthermore, some aspects of the results challenge the prevailing understanding of GCR transport. Notably, recurrent 27-day flux variations were observed during specific time intervals; intriguingly, these variations were more pronounced at higher particle energies, contrary to theoretical predictions that they would disappear at such energies. Additionally, the recurring electron-flux variations on shorter timescales serve as a robust observational constraint for models aiming to explain the time-dependent solar modulation of GCRs. The study also identified significant structures within the electron–proton hysteresis, corresponding to sharp flux fluctuations. These continuous daily electron data provide invaluable insights into the charge sign dependence of CRs throughout the 11-year solar cycle.

Discussing this manuscript, Du Toit Strauss and Engelbrecht [12] emphasized reproducing these precise measurements for GCR electrons and protons using solar modulation models. Such efforts can yield valuable insights into these particles' transportation mechanisms and advance knowledge in this field. Furthermore, accurate prediction of GCR flux and its associated radiation levels is crucial for ensuring the safety of human exploration in the solar system.

8.2.5 *Equivalent and effective dose estimation*

Measurements of absorbed doses alone using passive dosimeters are inadequate for studying biological effects or assessing radiation risk for astronauts. Dose equivalents must consider the entire spectrum of radiation distribution, including QFs and RBE of high-LET particles present in the SR environment. Therefore, CRD data can supplement absorbed dose measurements and provide a more comprehensive assessment of radiation levels at specific installation sites or areas.

In a manuscript published in 2023 by Chen X. and collaborators [13], it was reported that the calculation precision of the astronaut radiation dose induced by energetic CRs from outside the solar system has been improved. This improvement was achieved by utilizing data from astroparticle experiments, particularly the AMS-02 data. The paper presents a new calculation of the astronaut dose rate from GCRs in free space at 1 AU. To calculate the GCR spectra at 1 AU, a new 3D, time-dependent solar modulation model based on Parker's transport equation, initially developed by Song *et al.* [14], was employed. This model incorporates the recent LIS spectra and the measured data from the PAMELA and AMS-02 astroparticle experiments conducted between 2006 and 2019. The GCR spectra and the fluence-to-dose conversion coefficients were used as inputs to determine the dose-equivalent contribution.

Also, an example of the potential use of AMS-02 for implementing a radiation monitoring system in the external environment of the ISS was presented [15]. This type of system could be a valuable resource for planning spacewalks or missions involving human crews in low Earth orbit using a smart-system approach that combines, analyzes, and provides feedback and alarms based on inputs from instruments inside [16, 17] and outside the ISS and on satellites in different orbits.

8.2.5.1 *Comparison of absorbed dose ranges: astronauts in space missions vs. diagnostic and therapeutic procedures on Earth*

Medical procedures involving IR are standard in diagnostic imaging (e.g., X-rays and computed tomography scans) and cancer treatment (e.g., external beam radiotherapy and brachytherapy). These procedures aim to diagnose medical conditions or deliver targeted radiation doses to treat cancer while minimizing radiation exposure to surrounding healthy tissues. Health effects depend on the absorbed doses. The absorbed doses from diagnostic imaging procedures are generally low and unlikely to cause acute health effects. Absorbed dose ranges depend on the type of procedure: for diagnostic procedures, the absorbed doses for diagnostic imaging procedures vary depending on the type and complexity of the exam. For example, a typical chest X-ray may deliver an absorbed dose of 0.1–0.2 mSv, while a CT scan of the abdomen can range from 5 to 20 mSv or higher. However, repeated exposure over time may slightly increase the risk of developing cancer, particularly in sensitive tissues.

Therapeutic radiation treatments selectively target and destroy cancer cells while minimizing damage to healthy tissues. Side effects may occur depending on the treatment site, dose, and individual patient factors, but they are typically manageable and temporary.

In radiotherapeutic cancer treatment, absorbed doses can vary significantly depending on the type and stage of cancer and the treatment modality. For example, conventional external beam radiotherapy may deliver doses ranging from 1.8 to 2.0 Gy per fraction. In contrast, stereotactic radiosurgery (SRS) or stereobody radiotherapy (SBRT) may deliver higher doses per fraction (e.g., 10–20 Gy).

Space missions and medical procedures on Earth involve exposure to IR, albeit in vastly different environments and contexts. In this comparison, we will explore the absorbed dose ranges experienced by astronauts during space missions and those encountered by individuals undergoing diagnostic and therapeutic procedures on Earth. While both scenarios involve radiation exposure, the magnitude, duration, sources, and effects of radiation vary significantly between space exploration and medical applications.

Astronauts face unique challenges regarding radiation exposure during space missions due to the absence of Earth's protective atmosphere and magnetic field. They are exposed to various sources of IR, including GCRs, SPEs, and trapped radiation in the Van Allen belts.

Absorbed dose ranges vary according to the duration of the mission and place of permanence. Total mission dose: Astronauts on long-duration missions, such as those to the International Space Station (ISS) or future missions to Mars, can receive absorbed doses ranging from approximately 50 to 200 millisieverts (mSv) or more. Annual Limit: The annual radiation dose limit for astronauts is typically set at 1 millisievert (mSv) above the background radiation exposure on Earth, about 2–3 mSv per year.

The exposure of astronauts during space missions might lead to health effects, such as an increased cancer risk. Chronic exposure to cosmic radiation increases the risk of developing cancer, particularly in organs with high radiation sensitivity, such as the colon, stomach, and lungs. In an SPE or exposure to high doses of radiation, astronauts may experience symptoms of acute radiation sickness, including nausea, vomiting, fatigue, and potential damage to the bone marrow and gastrointestinal tract.

In conclusion, while space missions and medical procedures involve exposure to IR, there are significant differences in the sources, doses, and health effects experienced by astronauts compared to those on Earth. Astronauts face challenges related to chronic exposure to cosmic radiation during long-duration space missions, which may increase their risk of developing cancer over time. In contrast, medical procedures on Earth aim to provide diagnostic information or targeted cancer treatments while minimizing radiation exposure to patients and medical staff. The absorbed dose ranges for diagnostic and therapeutic procedures vary depending on the specific modality and clinical indication, with careful consideration given to optimizing the balance between diagnostic or therapeutic efficacy and radiation safety. Ultimately, both space agencies and medical professionals prioritize the health and well-being of individuals while harnessing the benefits of IR for exploration and healthcare.

8.2.6 *Ground- or space-based experiment definition*

Ad hoc measurements in the lunar gateway/lander or spacecraft biophysical laboratory are expected to further the knowledge of RBE and QFs for space missions. Space-based experiment setups can be identified and improved by replicating *ad hoc* experiments on Earth for relevant endpoints in preparation for future space missions. The possibility of conducting *ad hoc* experiments on Earth allows for overcoming uncertainties due to the limited number of subjects involved in space missions and paves the way to an era of lunar and Martian missions using more accurate risk models. CDR data are essential for validating GCR models, focusing on the 1–10 GeV/n range.

8.2.7 *Absorbed dose computation for space exposure scenarios*

MC codes can be implemented to calculate the absorbed dose and predict or describe the effects of GCR particles interacting with the cells, tissues, and organs of astronauts. These can be modeled as geometries with increasing details and complexities. The CRD data could be used as input data for the MC codes to determine the absorbed dose in the forecast exposure scenario (e.g., lunar gateway/lander or spacecraft).

8.2.8 *Other possible improvements for risk assessment*

Several aspects of IR-induced health effects still require investigation before undertaking long-duration space missions beyond Earth's magnetic protection. These areas could benefit significantly from the data collected by CRDs. Among the many topics in space radiobiology, one promising direction for improving risk models is the incorporation of non-targeted effects, such as the bystander effect, where cells or tissues not directly irradiated still exhibit damage due to signals from nearby irradiated cells [18].

As a representative example of using CRD data for improving dose–effect models, GCR proton fluxes measured by the AMS-02 detector using different energy bins were used to evaluate tumor prevalence (TP) versus exposition time for GCR protons [19, 20]. For this study, an R-script library has been developed to consider all relevant cell survival probability (CSP) models, which describe the relationship between radiation dose and the proportion of cells that survive for CPs suitable for SR.

Based on the multi-target single-hit model for protons, cellular TST was used to assess the TP regarding the hazard function (HF). The HF considers the probability of the event and the possible associated damage. The IDM mathematical core function depends on the charge number (Z), kinetic energy (E), and fluence (F). More details on this research will be provided in the following sections.

8.3 An Example of Enabling Research at INFN AMS Roma Sapienza: Dose–Effects Models for Space Radiobiology

After assessing potential improvements from the CRD data, we focused our research on the dose-effects models for space radiobiology. Enhancing this part will be crucial for the risk assessment of future space missions.

8.3.1 *Introduction to dose–effect models in medicine and public health*

Dose–effect relationships describe the correlation between the amount of exposure to a harmful agent (such as IR) and the resulting biological effects on the body. These relationships are fundamental in medicine and public health, as they inform decision-making processes regarding acceptable levels of exposure and guide the design of protective measures. In the context of radiation, the dose–effect relationship helps determine the threshold levels at which radiation exposure harms human health and identifies the types of biological damage it might cause, such as cancer, tissue damage, or genetic mutations. The dose–effect relationship serves as the foundation for radiation protection models. Understanding how different amounts and types of radiation influence the body is essential. The dose is typically measured in units such as sieverts (Sv) or gray (Gy), which consider both the energy deposited by the radiation and the biological effectiveness of the radiation type. Models based on this relationship are crucial for establishing guidelines for safe radiation levels in various settings, such as medical imaging, occupational exposure, and space missions.

Dose–effect models are pivotal in determining the optimal radiation doses for diagnostic procedures and cancer treatments. For instance, radiation therapy for cancer patients involves carefully calibrated doses to target and destroy cancer cells while minimizing harm to surrounding healthy tissue. Understanding the precise dose–response relationship allows healthcare professionals to tailor treatments for each patient, enhancing the therapeutic effects while reducing the risks of side effects. On the other hand, public health policies use these models to regulate environmental radiation exposure, thereby ensuring that the public remains protected from harmful radiation levels, whether from natural sources, such as CRs, or artificial sources, such as nuclear power plants. In public health, understanding dose–effect relationships is also key for managing potential radiation emergencies. These models help estimate the potential impact of radiation exposure on large populations in the event of accidents, allowing for the development of effective evacuation, treatment, and decontamination strategies.

8.3.2 *An overview of scientific literature on dose relationship for space radiobiology*

One of the main steps in starting the research on dose–effect models in space radiobiology consists of conducting and publishing an assessment of the existing literature [21]. To summarize the methodology and focus of a literature review on dose–effect models in space radiation, the study evaluates the development and validation of mathematical models that describe the relationship between radiation dose and biological effects. These models are informed by data from human studies, *in vitro* cellular experiments, and *in vivo* small animal research, providing insights into both clinical and subclinical effects observed during space missions.

Ground-based simulations using diagnostic or therapeutic radiology equipment allow researchers to recreate GCR exposure scenarios, leveraging similarities in radiation dosage and particle types to improve our understanding of the effects of space radiation. A notable area of focus is the study of non-targeted effects induced by secondary particles, which play a significant role in cancer risk at doses relevant to space environments. Emerging research also explores innovative countermeasures, such as hibernation, which could benefit radioprotection through induced hypothermia.

The review explicitly examines the acute and long-term adverse effects of space radiation, comparing them with outcomes observed from diagnostic and therapeutic applications on Earth. The study employed a systematic search strategy using PubMed/Medline, focusing on keywords related to space radiobiology and dose–effect models, restricting recent studies from the past decade (2010–2021). Abstracts were screened, and relevant studies underwent further review, with additional references identified through the bibliographies of selected papers.

Sixty-one papers were included, comprising 54 original studies and eight review articles. Data sources spanned spaceflight observations and ground-based experiments, with 24 papers using astronaut data and 37 focusing on simulations. Models described various health effects, including eye flashes, cataracts, central nervous system (CNS) impacts, cardiovascular disease (CVD), cancer, and biomarkers. Each model's reliability and research priority were assessed based on the robustness of the data and their implications for long-term space missions.

The findings highlight a diverse range of radiation-induced health risks, categorized into short-term effects (e.g., eye flashes), medium-term effects (e.g., CVD), and long-term effects (e.g., cancer), with timeframes ranging from immediate to several decades post-exposure. Despite advancements, the reliability of many models still needs to be improved, underscoring the need for continued research to support future interplanetary exploration. The review provides a foundation for understanding radiation risks and developing protective measures for astronauts.

Table 8.2 summarizes the results, and Fig. 8.2 shows our evaluation in terms of reliability and priority in investigations concerning the different aspects evaluated.

Table 8.2. Overview of the biological effects of radiation exposure from spaceflight and ground-based simulations, highlighting dose thresholds, dose ranges, and study prevalence.

Effect	Study Types	Dose Range/Threshold	#Papers
Eye Flashes	Spaceflight	LET >5–10 KeV/μm	4
Cataract	Spaceflight	Eight mSv	5
CNS	Ground/Simulations	100–200 mGy	11
CVD	Spaceflight	1,000 mGy	4
	Ground/Simulations	0.1–4,500 mGy	8
Cancer	Spaceflight	<100 mGy	2
	Ground/Simulations	<100 mGy	9
Biomarkers and Chromosomal Aberrations	Spaceflight	5–100 mGy	11
	Ground/Simulations	<10,000 mGy	4
Other Risks	Ground/Simulations	~2,000 mGy	2

Dose - Effect Models Overview
Evaluation

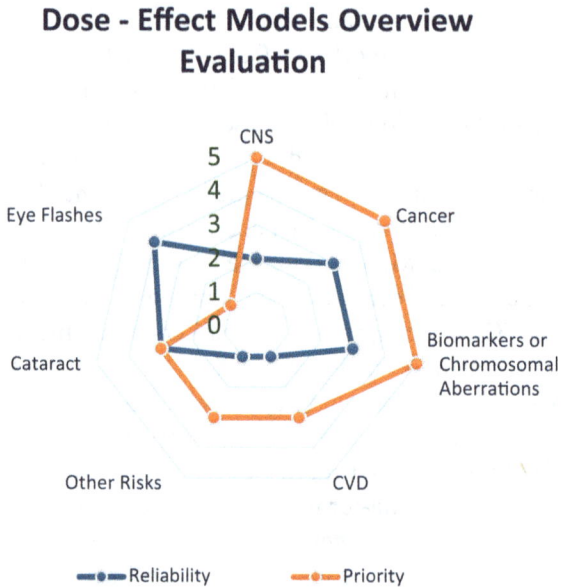

Fig. 8.2. This radar chart compares the reliability and priority of dose–effect models across different health effects associated with space radiation exposure. Reliability scores (blue line) assess the robustness of the models based on available data, statistical approaches, and an understanding of space radiation spectra. Priority scores (orange line) represent the significance of these effects for future research, emphasizing their potential impact on astronaut health during long-term missions. Effects such as CNS and cancer carry a high priority due to their implications for mission duration and quality of life, while other risks are assigned relatively lower priorities. The scoring system ranges from 1 (very low) to 5 (very high).

Source: Generated with licensed MS Excel by the authors.

This study highlights significant advancements in risk assessment capabilities, made possible by accessing information on energy ranges that were previously unexplored with such accuracy and precision. Furthermore, comprehensive species of cosmic radiation, ranging from elementary particles (electrons and protons) to light and heavy nuclei (from helium to iron and beyond), are now considered in the assessment, as they may pose concerns for space missions. To achieve this objective, once the exposure scenario for the space mission is established, which involves accurately identifying the mission details such as duration, destination, and the forecasted period to evaluate solar activity conditions that modulate GCR radiation components, the risk assessment will calculate potential damages associated with the space mission.

The table presents a synthesis of research on radiation-induced biological effects, drawing from both spaceflight and ground-based simulation studies. Effects are categorized based on the origin of the data: spaceflight, which reflects *in vivo* human or animal exposure in space, and ground/simulations, which emulate space radiation environments using particle accelerators or other laboratory setups or *in silico* calculations. "Dose Range/Threshold" indicates the dose or LET values studied, with ranges representing varying radiation conditions, from low (<100 mGy) to high doses (~2,000 mGy). Threshold values signify the minimum dose or LET required to observe a given effect. "#Papers" refers to the number of studies included in the review for each category, indicating research intensity [21].

8.3.3 *Using the AMS data: the NTE-DEM tool*

Following this analysis, we developed an *ad hoc* software tool (NTE-DEM) to investigate one of IR's promising but not yet understood effects, usually referred to as the non-targeted effect (NTE), which is relevant for space radiation. These effects include bystander signaling between irradiated and non-irradiated cells and genomic instability in cell progeny. NTEs are crucial for understanding health risks to astronauts, especially for long-term missions [19]. Figure 8.3 shows the basic concepts of IR-induced NTE and the bystander effect as a special case.

NTE-DEM integrates existing experimental data, CR fluences measured by the AMS, and CSP functions from the literature. The tool implements RStudio, with over 10,000 lines of code, including a main program and various libraries.

Initially, it has been used to model TP induced by proton radiation and validated with preclinical *in vivo* data on Harderian gland tumors and AMS proton flux data. Notably, the tool contains a library in which all the CSP functions described in Chapter 6 and used cells are coded. Figure 8.4 represents a dose–effects model tuned using the NTE-DEM tool [20].

Future extensions of NTE-DEM will include the following:

- incorporating experimental data for other diseases induced by space radiation (e.g., CNS effects, CVD);
- expanding the AMS dataset to analyze additional CR components (e.g., heavy nuclei, electrons);
- including radiobiological mathematical models for specific NTE mechanisms (e.g., intercellular communications);

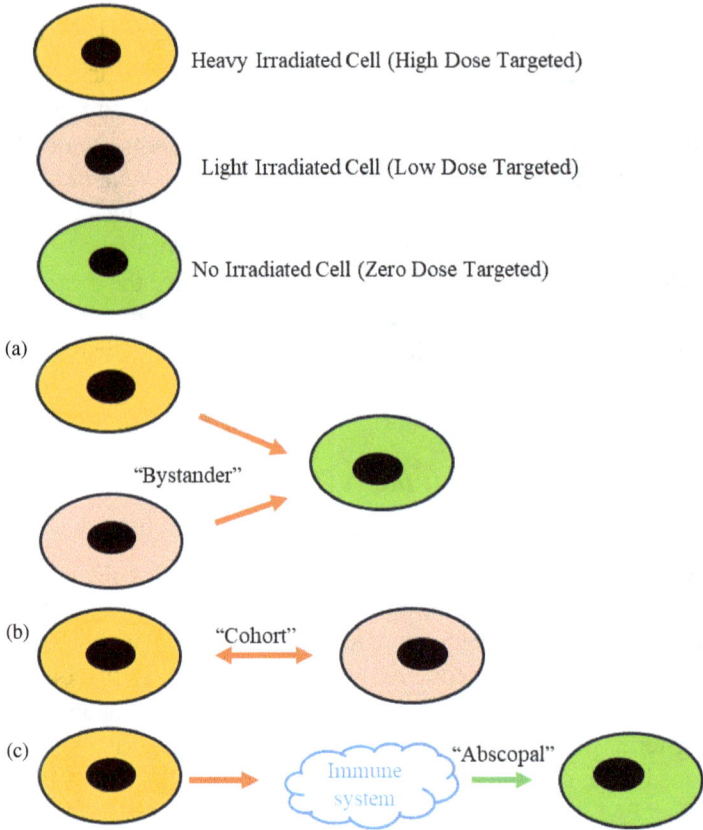

Fig. 8.3.   Radiation-induced non-target effects.

Local: (a) "Bystander" effect – refers to the impact on neighboring cells that are not directly irradiated but are affected through intercellular signaling mechanisms. (b) "Cohort" effect – describes effects observed in a population of cells sharing proximity or exposure conditions, with potential interactions influencing their collective response.

Distant: (c) "Abscopal" effect – refers to systemic effects occurring in tissues or organs far from the site of radiation exposure, often mediated by immune system activation or other systemic signaling pathways.

Source: Generated with licensed MS PowerPoint by the authors.

- applying AI to analyze datasets for risk assessments and health hazard forecasting;
- a database will be created for AI model training, validation, and testing, enabling precise predictions of space radiation health risks.

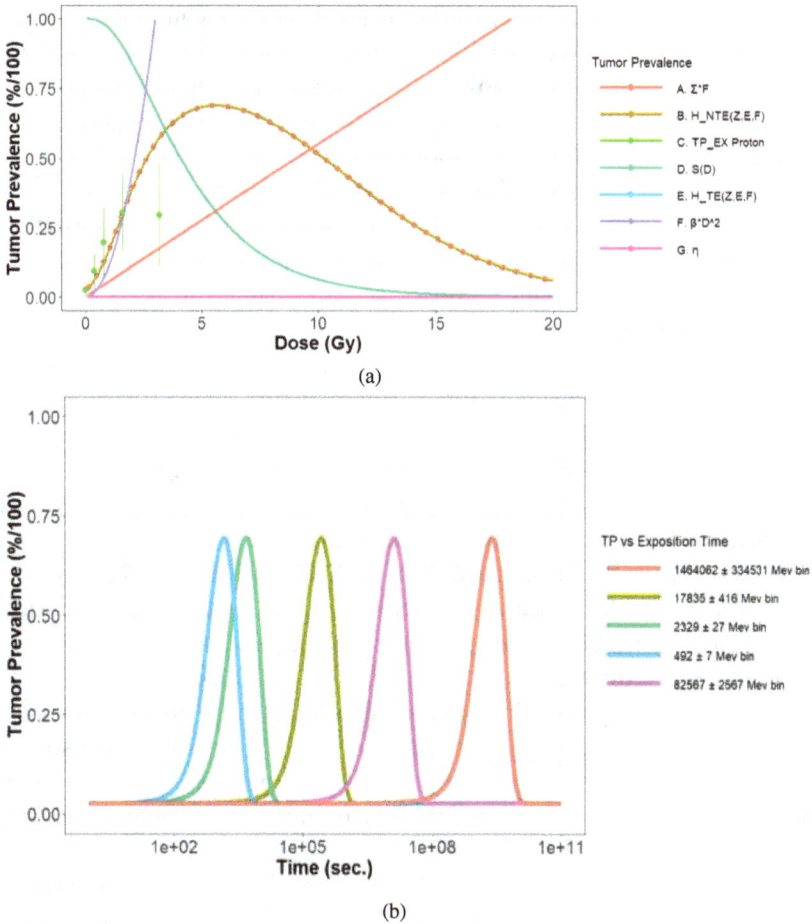

Fig. 8.4. (a) Dose–effect model for Harderian gland tumor generated using the NTE-DEM tool. (b) Applications of the model to evaluate the damage due to the proton expositions as measured by the AMS-02 detector as functions of time. Different ranges of energies are plotted.

Source: Generated with a licensed NTE-DEM tool in R scripting language by the authors.

8.4 Synergies in the Future

Research into the effects of IR in clinical and space contexts presents numerous opportunities for synergy, particularly when considering the nature of particles and their dangerous effects. Regarding particle interactions, protons play a central role in both fields. In clinical settings, proton therapy is widely used for its ability to target tumors with high precision while sparing surrounding healthy tissues. Similarly, protons are a dominant component of cosmic radiation in space, posing risks of deep tissue damage to astronauts. By studying dose deposition and biological effects, researchers can advance cancer therapies and improve our understanding of the health risks associated with prolonged space travel.

Electrons, commonly used in radiotherapy for treating superficial tumors, are also present in Earth's radiation belts, contributing to skin and ocular damage in astronauts. Collaborative research on tissue responses and the development of shielding techniques can benefit both clinical treatments and space mission safety. High-charge, high-energy ions, such as carbon, are critical in heavy ion therapy, which offers highly localized and precise radiation treatments. These ions are also a significant component of galactic CRs, known for causing dense ionization damage. Investigating DNA damage repair and cellular responses to these ions offers potential advancements in cancer therapy and the mitigation of SR effects.

Neutrons, produced as secondary radiation in proton therapy, also arise when space radiation interacts with spacecraft materials. Understanding the effects of neutron shielding and secondary radiation could optimize safety protocols for patients undergoing radiation therapy and astronauts in space. Photons, along with X-rays and gamma rays, constitute the cornerstone of conventional radiotherapy and are also generated as secondary radiation in space. Research on radioprotective agents and the effects of low-dose radiation is relevant to medical diagnostics and space travelers' long-term health.

Focusing on the dangerous effects, both fields share an interest in carcinogenesis. Clinical research emphasizes understanding cancer induction by therapeutic radiation, while space research investigates cancer risks associated with chronic low-dose radiation exposure during missions. Both fields can benefit from improved prevention and treatment strategies by exploring biomarkers for radiation-induced cancer, dose–response relationships, and individual susceptibility. CVD is another shared area of concern, with radiation exposure contributing to heart disease in patients receiving chest radiotherapy and endothelial damage in astronauts. Studying endothelial responses and developing protective interventions can address these risks in both settings.

The CNS is another critical area of synergy. In clinical research, the focus is on the cognitive decline and neuroinflammation associated with brain radiotherapy. In contrast, space research investigates the long-term

Table 8.3.   Possible synergies between clinical and space radiation research based on (a) particle type and (b) dangerous effects or diseases.

(a)

Particle Type	Clinical Research Focus	Space Radiation Research Focus	Potential Synergies
Protons	Used in proton therapy for cancer treatment to minimize off-target effects.	The dominant component of space radiation causes deep tissue damage.	Studying dose deposition and biological effects to improve cancer therapy and understand astronaut health risks.
Electrons	Used in radiotherapy for superficial tumors.	Present in Earth's radiation belts, leading to skin and ocular damage.	Developing shielding techniques and studying tissue responses for both treatment and astronaut protection.
High-*Z* Ions (e.g., Carbon)	Used in heavy ion therapy for highly localized and precise treatments.	Galactic cosmic rays (GCR), including heavy ions, cause dense ionization damage.	Researching DNA damage repair and cellular responses to improve clinical therapies and mitigate space radiation effects.
Neutrons	Secondary radiation in clinical settings, particularly in proton therapy.	Produced during space radiation interactions with spacecraft materials.	Understanding neutron shielding and secondary radiation effects to optimize safety in clinical and space environments.
Photons (X-Rays/Gamma)	The primary tool in conventional radiotherapy.	Generated as secondary radiation in space.	Advancing radioprotective agents and studying low-dose effects relevant to prolonged space missions and medical diagnostics.

(*Continued*)

Table 8.3. (*Continued*)

(b)

Effect	Clinical Research Focus	Space Radiation Research Focus	Potential Synergies
Cancer	Understanding cancer induction by therapeutic radiation.	Studying cancer risks from chronic low-dose radiation in space.	Investigating biomarkers for radiation-induced cancer, dose–response relationships, and individual susceptibility.
Cardiovascular Disease (CVD)	Radiation-induced heart disease from chest radiotherapy.	Radiation exposure contributes to endothelial damage and CVD risk.	Examining endothelial responses and developing interventions to prevent radiation-induced CVD in clinical and space contexts.
Central Nervous System (CNS)	Studying cognitive decline and neuroinflammation in brain radiotherapy.	Assessing long-term CNS effects of space radiation on astronauts.	Researching neuroprotection strategies and the mechanisms of radiation-induced neurotoxicity for improved treatments and space health management.
Acute Radiation Syndrome (ARS)	Rare in clinical settings but critical in accidental exposures.	Relevant in intense solar particle events during space missions.	Enhancing acute countermeasures like radioprotectors and therapeutics for medical emergencies and space scenarios.
Tissue Degeneration	Fibrosis and chronic damage in radiotherapy-treated tissues.	Degeneration from prolonged radiation exposure in space.	Exploring antifibrotic agents and regenerative therapies applicable to both fields.
Genomic Instability	Impact on secondary cancers and hereditary effects in patients.	Persistent genomic instability in astronaut cells.	Studying DNA repair pathways, epigenetics, and radiation-induced mutations for better prevention and monitoring strategies.

effects of space radiation on astronauts' CNS health. Collaborative efforts to understand the mechanisms of radiation-induced neurotoxicity could lead to neuroprotection strategies for both patients and astronauts. Acute radiation syndrome (ARS), though rare in clinical practice, is relevant in cases of accidental exposure and during intense solar particle events in space. Research aimed at enhancing countermeasures, such as radioprotectors and therapeutics, can improve responses to radiation emergencies in both contexts.

Tissue degeneration is another significant overlap. In clinical settings, chronic radiation exposure often results in fibrosis and long-term tissue damage, while prolonged space radiation exposure also contributes to similar degenerative processes. Investigating antifibrotic agents and regenerative therapies could yield solutions to mitigate these effects in both environments. Finally, genomic instability presents challenges in both clinical and space environments. In patients, it contributes to the risk of secondary cancers and potential hereditary effects. For astronauts, ongoing genomic instability can compromise long-term cellular health and increase susceptibility to radiation-related diseases. Advances in understanding DNA repair mechanisms, epigenetics, and radiation-induced mutations hold promise for enhancing prevention and monitoring strategies across both disciplines.

Through these shared challenges and opportunities, research in clinical ionizing and space radiation can benefit from collaborative efforts, advancing technology and medical interventions that improve human health on Earth and in space. Table 8.3 highlights some suggestions starting from the points of view of particles (a) and effects (b).

References

[1] Pruzan, P. (2016). *Research Methodology: The Aims, Practices and Ethics of Science*. Springer International Publishing, Switzerland. https://doi.org/10.1007/978-3-319-27167-5.

[2] ISECG. (2024). *Global Exploration Roadmap*. International Space Exploration Coordination Group, August 2024. https://www.globalspace-exploration.org/wp-content/isecg/GER2024.pdf.

[3] Bartoloni, A., Della Gala, G., Paolani, G., Santoro, M., Strigari, L., Strolin, S., and Guracho, A. N. (2022). High energy physics astroparticle experiments to improve the radiation health risk assessment in space missions. *Proceedings of Science*, EPSHEP2021. https://doi.org/10.22323/1.398.0106.

[4] Bartoloni, A., Della Gala, G., Paolani, G., Santoro, M., Strigari, L., Strolin, S., and Guracho, A. N. (2022). Astroparticle experiments to improve the radiation health risk assessment for humans in space missions. *Proceedings of the International Astronautical Congress*, IAC2022.

[5] Bartoloni, A., Della Gala, G., Paolani, G., Santoro, M., Strigari, L., Strolin, S., and Guracho, A. N. (2022). Dose-effects models for space radiobiology: An overview on dose-effect relationship. *Proceedings of the International Astronautical Congress*, IAC2022.

[6] Bartoloni, A. and Strigari, L. (2021). Can high energy particle detectors be used for improving risk models in space radiobiology? *Proceedings of the Global Exploration Forum*, GLEX2021.

[7] Badhwar, G. D., O'Neill, P. M., and Troung, A. G. (2001). Galactic cosmic radiation environment models. *AIP Conference Proceedings*, 552, 1179. https://doi.org/10.1063/1.1358069.

[8] Slaba, T. C. and Whitman, K. (2020). The Badhwar-O'Neill 2020 GCR model. *Space Weather*, 18, e2020SW002456. https://doi.org/10.1029/2020SW002456.

[9] Väisänen, P., Usoskin, I., Kähkönen, R., Koldobskiy, S., and Mursula, K. (2023). Revised reconstruction of the heliospheric modulation potential for 1964–2022. *Journal of Geophysical Research: Space Physics*, 128, e2023JA031352. https://doi.org/10.1029/2023JA031352.

[10] Koldobskiy, S. A., Bindi, V., Corti, C., Kovaltsov, G., and Usoskin, I. (2019). The neutron monitor yield function was validated using data from the AMS-02 experiment 2011–2017. *Journal of Geophysical Research: Space Physics*, 124(4), 2367–2379. https://doi.org/10.1029/2018JA026340.

[11] Aguilar, M., *et al.* (AMS Collaboration). (2023). Temporal Structures in Electron Spectra and Charge Sign Effects in Galactic Cosmic Rays. *Physical Review Letters*, 130, 161001. https://doi.org/10.1103/PhysRevLett.130.161001.

[12] Du Toit Strauss, R., and Engelbrecht, N. E. (2023). Disentangling the Sun's Impact on CRs. *Physics*, 16, 62.

[13] Chen, X., Xu, S., Song, X., Huo, R., and Luo, X. (2023). Astronaut Radiation Dose Calculation with a New Galactic Cosmic Ray Model and the AMS-02 Data. University of Jinan, UJN, Jinan, China, Shandong Institute of Advanced Technology, SDIAT, Jinan, China, 03 April 2023.

[14] Song, X., Luo, X., Potgieter, M. S., Liu, X. M., and Geng, Z. (2021). A Numerical Study of the Solar Modulation of Galactic Protons and Helium from 2006 to 2017. *The Astrophysical Journal Supplement Series*, 257(48), 1–13. https://doi.org/10.3847/1538-4365/ac281c.

[15] Faldi, F. (2022). Real time monitoring of the radiation environment on the ISS with AMS-02. *IL Nuovo Cimento C*, 45, 1–8. https://doi.org/10.1393/ncc/i2022-22079-6.

[16] Morone, M. C., Berucci, C., Cipollone, P., De Donato, C., Di Fino, L., Iannilli, M., La Tessa, Manea, C., Masciantonio, G., Messi, R., *et al.* (2019). A compact Time-Of-Flight detector for radiation measurements in a space habitat: LIDAL–ALTEA. *Nuclear Instruments and Methods in Physics Research Section A*, 936, 222–223. https://doi.org/10.1016/j. nima.2018.09.139.

[17] Berger, T. (2008). Radiation dosimetry onboard the International Space Station ISS. *Zeitschrift für Medizinische Physik*, 18, 265–275. https://doi. org/10.1016/j.zemedi.2008.06.014.

[18] Cucinotta, F. A. and Cacao, E. (2017). Non-targeted effects models predict significantly higher mars mission cancer risk than targeted effects models. *Scientific Reports*, 7, 1832. https://doi.org/10.1038/s41598-017-02087-3.

[19] Guracho, A. N., *et al.* (2022). Target effects vs. non-target effects in estimating the carcinogenic risk due to galactic CRS in exploratory space missions. *Proceedings of the International Astronautical Conference 2022*, IAC-22-A1.5.70998, Paris, September 2022.

[20] Guracho, A. N., Strigari, L., Della Gala, G., Paolani, G., Santoro, M., Strolin, S., and Bartoloni, A. (2023). Space radiation-induced bystander effect in estimating the carcinogenic risk due to galactic cosmic rays. *Journal of Mechanics in Medicine and Biology*. https://doi.org/10.1142/ S0219519423400237.

[21] Strigari, L., Strolin, S., Morganti, A. G., and Bartoloni, A. (2021). Dose-Effects Models for space radiobiology: An overview on dose-effect relationships. *Frontiers in Public Health*, 9, 733337. https://doi.org/10.3389/ fpubh.2021.733337.

Index